How to Get a Job in Television

Elsa Sharp

A & C BLACK • LONDON

First published 2009

1 3 5 7 9 10 8 6 4 2

A & C Black Publishers Limited
36 Soho Square
London W1D 3QY
www.acblack.com

Copyright © Elsa Sharp

ISBN: 978 1 408 10129 2

Original design by Fiona Pike,
Pike Design, Winchester

Typeset by RefineCatch Limited, Bungay, Suffolk
Printed and bound in Great Britain by
CPI Cox & Wyman, Reading, RG1 8EX

This book is produced using paper that is made from wood grown in managed, sustainable forests. It is natural, renewable and recyclable. The logging and manufacturing processes conform to the environmental regulations of the country of origin.

Contents

Acknowledgements

This book is dedicated to the memory of my wonderful mother, Giuliana, who was so creative and an avid reader.

Acknowledgements and thanks go to the following for sharing their experience, knowledge and contributing to this book. Alan Sharp (my dad), Alice Dudley, Andrew O'Connor, Andrew Chaplin, Alex Marengo, Anna Blue, Anna Keel, Anna Richardson, Ben Barrett, The British Library, Charles Martin, Conrad Green, Claire Richards, Daisy Goodwin, David Minchin, Deborah Kidd, Dominic Crofts, Eiran Jones, Emily Gale, Emily Shanklin, Fiona Chesterton, Glen Barnard, Grant Mansfield, Harjeet Chhokar, Helen Beaumont, Jenny Popplewell, John Adams, Jon Crisp, Jo Taylor, Julia Dodd, Julia Waring, Kate Phillips, Katie Rawcliffe, Liz Mills, Louise Mason, Lucy Reece, Matt Born, Michelle Matherson, Moray Coulter, Nick Holt, Peter Bazalgette, Rachel Roberts, Sam Moon, Simon Warrington, Sophie Ardern, Susannah Haley, Richard Drew, Richard Hopkins, Ruth Wrigley, Tim Hincks, Victoria Ashbourne.

Thanks also go to the researchers who helped me with the book: Alice Hope, Doug Garside and Puja Verma for transcribing some of the book's interviews. Anna Symons and Lizzie Moore for doing some of the interviews and transcribing them. I hope the experience was useful!

Thank you to my husband Mark for your advice and my little girls Ava and Mina who let me sneak away to write in peace.

Thank you Jenny Ridout, my commissioning editor at A&C Black, for giving the chance to write the book and for your faith, help and support.

Foreword

Peter Bazalgette

Nothing delights me more than assisting young people who really want to enter the TV industry, who closely watch and love the medium, who have researched some of the production companies which make the shows and have already managed to get some media experience while at college. Nothing irritates me more, on the other hand, than being introduced to the gormless offspring of acquaintances who have emerged from college with little to recommend them and, in the absence of any other ideas, 'think I might go into TV.'

The previous paragraph may make me sound like an unsympathetic swine. But in fact I love helping ambitious graduates. I just want to see that they have already advertised their media ambitions in some practical way. Now that everyone can make their own content and distribute it on the likes of YouTube there are so many ways of demonstrating your interest in media.

The great thing about this book is that it is packed with pertinent advice from current practitioners – people who know what working in television today is all about. And it is, indeed, a great time to be entering our business. Let me explain why. When I went into television, in the 1970s, there were pretty well just two employers: the BBC and ITV. They made the shows and distributed them via the three channels they owned. In the BBC, if you behaved yourself, you could climb a corporate ladder and in four years you'd get your own office. To begin with it would have false curtains, but in six years you might even get real ones. Today, with tens of broadcasters owning hundreds of channels, with more than a thousand independent producers, not to mention new digital content houses, the choice and opportunities are far superior. And you have the chance to make decisions about your own career path rather than have them made for you by some remote personnel department. The fluid, freelance world may seem perilous, but to resourceful, self-confident individuals it is a boon.

One characteristic, I'd suggest, that all content creators ought to have is the desire to entertain, to capture people's emotions. This holds true whether you work in news, sport, game shows, reality, documentary, drama or animation. You have something to say and you want people to listen.

And good luck – if you have the fortitude to knock on enough doors you will always succeed.

Peter Bazalgette
TV format creator and former independent producer

Preface

Welcome to *How To Get A Job In Television*, a careers guide to help you get a foothold and create a career path in television production. If you want to work in TV I hope this book will be an entertaining and useful guide.

I've been working in TV as a freelancer since 1994 and I'd like to give you an invaluable insight into the challenges of working in the industry and a few tips on how to get on using my own experience and talking to successful TV professionals. There is no other guide like this.

TV is a notoriously difficult industry to get into and to progress within, there is no set career path and more people want to be in TV than any other career.

This is where *How To Get A Job In Television* can help.

This book is aimed at graduates, people in other careers wanting a change, and new entrants, such as runners, junior researchers and researchers wishing to develop and further their careers in the industry.

My aim is to give everyone information that is usually only gained through experience and observation. This handbook is packed with useful tips and information that you would not otherwise be privy to. There are six chapters full of advice and anecdotes about finding a job and sustaining a career in TV. It explains how to find work experience, look for work, make and develop contacts, write a CV and get your first job. It outlines what's involved in different job roles and the skills required to do them, and how to progress and stay employed.

I reveal the essential skills required to be a researcher, assistant producer, producer, self-shooting producer/director, director and series producer and explain the different career routes and choices to make to reach these goals based on my own experience as a TV researcher, my knowledge as a series producer and a recruitment executive, and through interviews with those in the know.

A significant part of the book gives training advice to researchers: from how to find locations to dealing with difficult contributors. The researcher is a key role in television and the starting point for a career which could lead to producing and/or directing, executive producing and commissioning. The researcher is responsible for coming up with ideas, angles and stories, finding the content, characters, locations and setting up the programme and delivering the producer's vision and demands. Training and developing researching talent will create the programme makers of the future.

This guide draws on a range of sources from people working within the industry at every level to training bodies and HR executives within independent production companies. I've interviewed and profiled key TV executives: people who have forged successful TV careers, who reveal how they entered the industry, what qualities and attributes have helped them and what they consider to be essential key skills.

It includes contributions from across the board – runners, freelance researchers, managing directors of independent production companies and heads of creative talent. They have all been generous with their time and gave honest advice with candour.

This is not an academic text but it is packed with practical tips gained from hard earned experience. It is certainly something I wish I had been able to buy when I was starting out. There were no books, no access to the internet and no company websites back then!

You think you want to work in television? Well, here's where you make a start.

PROFILE

Richard Hopkins, managing director of Fever Media

 In his three years as creative head of format entertainment at the BBC, Richard was responsible for developing, pitching and producing a number of hit shows including *Strictly Come Dancing* which has been sold to over 31 territories worldwide.

He had a key role in the success of *Strictly Come Dancing* as an international hit, winning the Rose D'Or, Broadcast Award 2005, RTS 2005, Tric Award 2005 and 2006, TV Quick 2005 and 2006 as well as selling it to ABC in the USA, where it has been nominated for several awards including an Emmy.

Fever was launched in April 2006 by Richard and Emmy-award-winning producer David Mortimer. They are two of the UK's most respected TV executives, show runners and entertainment format creators. In its first two years Fever had a raft of new commissions.

Production credits

April 2008 – at the time of writing (spring '09) Managing Director of Fever Media
Responsible for developing and executive producing:
Fortune: Million Pound Giveaway (ITV1), *The People's Quiz* (BBC1), *No Place Like Home* (ITV1), *Murder Most Famous* (BBC2), *Britain's Bravest* (five), *Find Me The Face* (BBC 3)

2003–2006 Creative Head of Format Entertainment
BBC Television
Strictly Come Dancing (BBC1), *Strictly Dance Fever* (BBC1), *It Takes Two* (BBC2), *Hard Spell* (BBC1), *Star Spell* (BBC1), *We'll Meet Again* (BBC1), *The House of Tiny Tearaways* (BBC3)

Executive Producer at BBC Entertainment

Mastermind, Junior Mastermind, The Weakest Link, National Lottery Jet Set, A Question Of Sport, Sudo-Q, Facing The Music, Restored To Glory, Traitor, The Sack Race, Come and Have A Go, Wright Around The World, Dear Father Christmas, The Generation Fame, Didn't They Do Well and *Never Mind The Fullstops* (BBC4)

2002–2003 Executive Producer, *Fame Academy* and *Comic Relief Does Fame Academy* Endemol UK

2002–2003 Executive Producer, *Fear Factor.* Endemol UK for Sky 1

2001–2002 Producer, *The Runner* (Pilot) ABC, Los Angeles

2001 Endemol UK Development Executive

2000–2001 Executive Producer, *Big Breakfast* Planet 24

2000 Live Series Producer, *Big Brother* (Series 1) Endemol UK

1999–2000 Series Producer, *11 O'Clock Show* Talkback

1999 Live Editor, *Big Breakfast* Planet 24

1988–1999 Series Producer, *Baby Baby* Wall To Wall

Planet 24 Senior Producer, Planet USA

Producer/Director, *Hotel Babylon* (BBC1), *The Big Breakfast, Gaytime, The Word* Planet 24

Various producer positions at radio stations: Kiss 100 (Emap), BBC Worldwide, Kiss FM (Pirate), Sky TV, Power FM, Sunshine Radio (France)

What does your job involve?

My day-to-day job is a combination of things; managing and controlling the finance of the company, working with the development team to come up with new ideas to pitch to various channels, and working with the production teams with ideas that have been commissioned to make them successful shows. I'm doing this both in the UK and in the US. It's a constant and very tricky process when you're running a small company, either because things aren't going as well as you want and you're struggling to create new business or things *are* going well and you're struggling to control new business.

How did you get into television?

I did work experience at Sky when it was a small company and at the same time I was working at Kiss FM which was a pirate radio station. I started working there during my university years and continued there afterwards. It was trying to win a legal license and I started doing a programme called *The Word*, which was a spoken word programme to give them a bit more credibility above and beyond just playing records.

I worked at a number of stations before finally going back to work on the by then legal Kiss. I became a producer on the day-time shows. I was looking for a breakfast presenter and I went to meet Chris Evans, who I used to listen to on GLR.

He said he would love to come and work at Kiss except he had just done a pilot show called *The Big Breakfast*. I then saw an advert for a researcher for the show in *The Guardian*. I applied and got Chris to write my reference. I got quite a good reference because we had become friends. I was interviewed and got the job. So I went from day-time producer at Kiss, a spoilt and easy job; spoilt by the record companies, and easy because it's producing music radio shows, which are significantly easier than television shows, to working night and day on *The Big Breakfast*, quite literally. But I learned to love it. I worked my way up through various roles, it was a sort of 'university of television' working on a big show like that because you move up quickly, you get to work on everything and you learn a lot quickly.

I worked as an AP on the features section of the show, as an AP on the outside broadcasts and as an AP on *Zig and Zag* in the studio, learning all sorts of aspects on that one show. I then worked as a producer.

What was your big break – meeting Chris Evans?

Yes. He used to invite an audience down to see his shows. I went to see it and I spoke to him afterwards. As soon as my wife Cecile started speaking he said, 'Oh, I really like your accent, why don't you come on my show?' She ended up doing little features on his show and we ended up becoming friends. When that job opportunity came up on *The Big Breakfast* it helped. It wasn't nepotism as such. I didn't ring him up and say, 'Oh, can I have the job?' I applied for the job as anyone else would, but I put him down as a reference because he knew me.

Did you have a game plan?

Not really because I'd never thought I wanted to work in the TV industry particularly, it just suddenly seemed like it wasn't a bad idea. I remember wanting to work on the music features just because I wanted to get free CDs! It was as trivial as that really!

What qualities have helped your succeed?

Being able to write. Human skills are very important for television, it's about getting on with your team, keeping your immediate boss happy, dealing with talent, creativity; all the time you've got to be thinking of better ways of doing something than any other show that's been on, or better ways of improving the show that you're producing. They might be tiny differences but cumulatively they'll make a show much better. Sometimes you notice how a show has changed hands and got worse, or better, and you're not quite sure what's changed but you can just feel the quality going out, and that's normally from the quality of a team, not just an individual.

What's been your best moment in TV so far?

Craig, the winner of *Big Brother 1*, coming out of the house when I was live series producer. I felt like that character in the *Truman Show*, controlling the show and if I stopped speaking the show would look really awful. You feel like you've really earned your money with that feeling! No one had experienced *Big Brother* before and Craig had been taken into the hearts of the nation. He came out without us really knowing what he was going to do. All the fireworks were going off in the background and he gave his £70,000 prize money to a Down's syndrome kid, which sounds like the most cheesily constructed reality TV moment, but it wasn't. It was just something he did and was incredibly moving at the time. I tried to talk to Davina but I was so choked up I couldn't actually speak for a few seconds. Not only was the show concluding that I had contributed a massive amount to it but it concluded in a way that you couldn't have scripted. So that was bigger than the other moment.

What was the worst thing?
Being booted out of somewhere. It's never nice. It happened to me once, although I always claim I resigned. You feel at that moment that you'll never work again and it's incredibly upsetting. But you learn a lot from something like that happening. You learn more from some failures than you do from having no failures at all.

And *Fame Academy* not doing well. Before it went on air, we were all making up our jokes and tabloid headlines like, 'Oh, what a shame academy' and 'lame academy'. It became successful eventually but it started off dismally and got slated in the press because it was a BBC reality show. That was really tough, the tabloids being chucked at you all the time, it was immense. But you learn an awful lot from it, you realise the power of the tabloids. Once they don't like you it's an incredible upwards struggle, it's a growing experience really.

Chapter One
The bad, the good and the ugly and me

The bad: TV – are you tough enough?

I could start this book with a eulogy about how lucrative, fantastic and fun it is to work in TV. It is. But let's start with a few home truths. If you want to work in TV you need to know what you're getting into. It's not a career for the faint hearted.

Satirical BBC 4 TV show *Charlie Brooker's Screenwipe* makes a few well-placed observations. The episode, *A Career in Telly*, is bitingly funny with more than just a modicum of truth. It starts:

'Ooh, look it's you, a fresh faced nigh on foetal 21 year old finishing your media studies degree with an earnest documentary about a local homeless man called Billy. It's well received and you send your CV to 10 million different production companies and finally get a reply from a company offering you a job as a runner – but it's a start . . .' The letter the innocent young hopeful is holding says, 'Please be prepared to sell your soul'.

Sometimes your ideals, values, family and friends can go out the window in pursuit of a challenging career in TV.

'Television is full of very entertaining, very exciting and very fun interesting people,' says Kate Phillips, head of development at BBC Entertainment. Yes, some of the most clever, gifted, funny and creative individuals you'll ever meet work in TV but there are also some pretty horrendous people too. 'It's also full of egos, so if you're sensitive I really don't think you'll last that long,' she adds.

'The most difficult thing about working in TV is probably the team you find yourself working with,' says assistant producer Harjeet Chhoker. 'Despite what people say about TV, it is a very hard industry to work in, with long hours, stressful situations and some of the most unglamorous situations you'll ever find yourself in; you'll be asked to do things you'll wish you didn't have to. Therefore, having a good team to work with is vital. You could be doing the worst programme in the world, but if you have a good team then it's not that bad. But there are times when you end up working for people who are rude, unreasonable, manipulative, discouraging, easily stressed and under pressure. You can work on the most resourced, big budgeted show, like *Big Brother*, and it can still be hell!'

Jo Taylor, talent manager at Channel 4 agrees, 'It's a very selfish industry and it's very ruthless. It doesn't suffer fools gladly or tolerate weakness. If someone's not able to do their job properly it's so incestuous that it gets around the industry really quickly. It's a very superficial industry. If you have a hit, everyone wants to know you, if not . . .'

Not only do you need to be tough to work in TV, you need to be absolutely driven to succeed. It's an industry that's notoriously difficult to get into and you're competing against hundreds of other clamouring wannabes. Because it's perceived to be so glamorous many, many people want to work in TV.

'Before you set out you should know that the competition is extraordinary. There are many more candidates trying to get into the television industry than there are jobs,' says Julia Waring, head of creative talent at RDF Television. 'You need to be very realistic.'

Internet TV website Production Base is *the* online job network for TV production. It's the most popular and easily accessed website and most freelancers use it to find work. It's primarily a marketplace for hiring talent where freelancers can post their CVs and details of their availability for TV production companies to view. Production companies also use it to advertise jobs.

According to statistics from Production Base (2006) there are 70,000 researchers currently working in TV, but on any one day there may be 5,000 subscribing freelancers but only 250 jobs posted. In these recent lean times of recession 100 applicants could be chasing just one job and there are hardly any jobs advertised at all.

Getting into TV is hard; it is all about who you know. According to Skillset (the skills sector council for the creative industries in Britain) only 27 per cent of people working in the industry heard about their most recent job through a traditional recruitment route such as an advertisement; 70 per cent rely on contacts to get a foothold. 'Word of mouth is very important in this business,' says Richard Hopkins, of Fever Media, 'which makes it difficult for outsiders to break into. Few jobs are advertised, it's about contacts.'

'You get jobs by knowing people, not necessarily intimate friends but the last person you worked for may take you on or recommend you to somebody else. In fact, the pool of good talented people is very small. That's the real horrible truth of it,' says Julia Waring.

You get work by hearing of jobs – not easy if you don't know anyone, as people tend to work with people they've worked with before. But if you're hungry, tenacious and good enough, once you are in you can progress fast. If you get along with people you'll hear about work through them and will be recommended to others. 'If they like you then they'll work with you again,' says Daisy Goodwin, managing director of Silver River. 'If I like someone I would tend to give them lots of breaks.'

'If somebody does a really good job then you are more likely to ask them back again,' says Julia Waring. 'You are always going to go with someone who's been recommended strongly rather than a new person. You are always going to go with the familiar and ask the same people back again.'

But making contacts and being good does not always ensure continuous work in an increasingly competitive freelance market of shrinking budgets, uncertainty and cut backs. Good people can remain out of work for long periods too.

'The downside of working in television is the instability, insecurity and the knowledge that however right or however good you are, it might not be enough. It's seen as a labour of love,' says Moray Coulter, production and talent executive at ITV Productions. 'However committed you are that's not necessarily the thing that's going to make your career shine.'

TV is a mostly freelance industry. It's unusual to stay with one company and become a contracted member of staff. You'll probably be on a short contract with no job security, no sick or holiday pay. You might be unemployed for weeks between contracts or go from one burn-out job to another without respite.

'Changing teams all the time in a predominantly freelance community is quite a difficult discipline,' says Richard Hopkins. 'I don't suppose many people in most professions will change the people they work with as much as happens in television.' This can sometimes work to your advantage – if you find someone difficult to get along with then you don't have to work with them for very long but it's important to get along with everyone and fit in which is not always very easy.

As time goes on and you work in the same genre you'll probably come across and work with the same people time and time again. You'll work in new production companies where you know people from previous productions. They will become friends as well as colleagues; as you progress up the ladder you'll be in a position to pick your own teams, colleagues and associates. And if you impress your producer you might be taken from job to job with them.

It's either feast or famine being a freelancer. In recent years with personnel changes at different channels and a down turn in advertising, not to mention a global recession, there are more good people competing for the same jobs which means more freelancers are out of work than ever. 2007 was a particularly bad year. 'The industry was in a state of flux with controllers, heads of departments and commissioners changing and things being restructured,' said Michelle Matherson, talent executive at BBC Factual. Some freelancers worked regularly but for many it was grim and many left the industry. Things have become worse in the recession with cutbacks and budget cuts at all broadcasters.

I've been a freelancer in TV since 1994 and have worked for 15 different production companies – I've been invited back to several a few times and worked at Planet 24 on and off for over five years in production. The longest I was out of work was three months. On average I might have a break of three to four weeks between productions. As a single woman in my 20s I didn't mind the long hours. To recuperate I'd take holidays after a stint on a production – despite the fear, between jobs, that I might never work again. This is a common feeling whatever level you are at.

Ruth Wrigley has had a successful career in television as a freelancer for over 20 years. She was the founding series editor on *The Big Breakfast*, the executive producer on the award winning first *Big Brother* and won a BAFTA for *How Do You Solve A Problem Like Maria?* She admits that in between jobs she experiences the same feeling of apprehension. 'I always think I'll never work again. That's what's so horrible about telly, it's like playing musical chairs. People work, work, work and suddenly they bring someone else in and the in crowd's the out crowd and vice versa.'

Being a freelancer is precarious and there can be long periods of unemployment. 'I was once out of work for five weeks when I finished my first AP job,' says assistant producer Harjeet Chhokar. 'I went from being a very experienced researcher to being a very inexperienced AP. Companies wouldn't give me a job because they felt I needed more experience as an AP. Eventually a small company which was making a series for ITV2, which didn't have a big budget, took a chance on me and I was AP for the whole four-part series.'

It is also full of dilemmas – do you take the first job that comes along or do you wait for a show you really want to make that you might not necessarily be offered? The important thing to remember is that it's all about making strategic decisions, grasping opportunities as they arise and thinking about whether the project will suit you as an individual. 'Just be true to yourself, that's the thing,' says Ruth Wrigley.

It's important to hold your nerve if you do turn something down and to have faith that the right thing will turn up – eventually. But this unpredictability doesn't suit everybody.

The good

I've had the most incredible experiences in a career that has given me unique access to people and stories, and has allowed me to work with some of the most talented TV people in the country. I've produced several comedy game shows that have been fun to make and been hilarious to write and edit. I've worked with talented writers like Lucy Porter, Nick Hildred, Robin Ince, Phil Nice and Dave Cohen. I've cried with laughter filming a hidden camera show in towns and cities across Britain, making surprise hits on unsuspecting members of the public.

I've met and interviewed a wide range of celebrities from Simon Cowell, Ant and Dec to Michael Caine and Robbie Williams, as well as charity campaigners, obsessive collectors and experts of subjects I could never master.

TV allows you to enter worlds and to ask questions that people would like to ask but never get the opportunity. 'It brings privileged access,' says Moray Coulter. 'You've got a free path into other people's lives, other organisations' lives. You get a unique view of the world that not many other jobs do.'

'I love the fact that you get inside people's lives in a way that you could never do in any other job. It's amazing that you can ask people anything,' says Liz Mills, managing director of Top TV Academy, a training and recruitment agency. 'Because you wear the TV hat you can go behind the scenes and go to all sorts of places.

TV is exciting and varied and it's never boring. 'Every day is different,' says Grant Mansfield, chairman RDF Media Group Content. 'No two days are the same. The truth about television is that it's enormous fun and collaborative, which I really like. People get well paid and what they make is watched by lots of people. People do work long hours and it is a ruthless industry but it gives you a grandstand view.'

'When I was working my way up it was three months in one place, six months in another, really good fun shifting and moving all the time. When you're doing a weekly live show the adrenalin is pumping; you're aiming for one show and then working on the next one.'

TV is full of bright, inventive, inspiring people. It's irreverent and creative. And it is constantly stimulating – there is nothing like the buzz of coming up with ideas for a programme and seeing them go out on air for thousands of people to watch and talk about.

'Nothing, nothing beats seeing the final product and knowing all the tears, blood and sweat was worth it,' says Harjeet Chhokar. 'Some of the best times of my life have been whilst I have been working. I honestly think that it is a privilege to do what we do. I love working in TV and the thrill of seeing your show on the screen or hearing people talk about it is great.'

'You have to live every day of your working life determined to enjoy it. But never underestimate how easy it is to fail in television,' says Richard Hopkins. 'The fear of failure drives you and the absolute desire to get the next big hit is what drive you on, like a sad addiction!'

If you can cope well under pressure, are happy to sacrifice your friends and home life at times, weather the storms of freelancing, suffer constant rejection and compete for jobs with your former colleagues, television can be the most satisfying career.

It's interesting, entertaining and absorbing, giving you the most incredible and unique access to real stories of ordinary and extra-ordinary people, celebrities, experts, heroes and villains. It's also very well paid especially when you reach the top with salaries paid to executive producers in the six figure bracket with bonuses, company shares and a percentage of the formats and programmes they generate on top.

The ugly – how do you plot the right career path?
The right job doesn't always come along when it suits you – luck, chance and being available at the right time all have a major part is play. This could mean not working for a few weeks and waiting or taking something that's less appealing because you need to

financially. It's not always easy to turn down work because it's the wrong type of show or the wrong rate if you have rent or a mortgage to pay, but if you want to make a name in a certain genre you must have the nerve and the finances to hold out for the right job.

If you want to work on a certain type of programme or genre it's best to try to concentrate on being continuously employed in that field, but it can be difficult. You won't get a job without the right experience but how do you get that experience if no one is willing to take a risk on you?

Working in TV at any level isn't always easy and is never straightforward. If you do manage to pull off the impossible you might not always get the recognition you deserve or even a thank you – and someone else is always ready to claim the credit. You have to be calm under pressure, cheerful, tenacious and resourceful even when the going is tough and you're tired beyond belief.

With more broadcast TV channels than ever and advertising revenue being diluted between these and internet websites, budgets are getting tighter, schedules and teams are getting smaller, making it harder for those who are making the programme, forcing them to do a job that would have been shared between two or three people. It is quite common for producer directors to shoot, direct, produce, do the sound and even edit their own programmes. In the past they would have had the support of a cameraman, soundman, researcher and editor.

When you start working in TV it can be extremely stressful, delivering seemingly impossible programmes to unworkable deadlines, schedules and budgets.

'TV can be a nightmare, where you can find yourself still at work at 4am, not having slept or eaten and you can't remember the last time you saw your friends or family,' says Harjeet Chokker. 'It's full of egos and quite nasty people who are not afraid to shout and make others cry. This is when you have to remember that TV is a great industry to work in'.

You have to be mentally and physically tough, confident and focused. 'You've got to be able to work well under pressure because inevitably there is going to be a deadline at some point when a show is going to go on air, whether it is pre-recorded or live, there's an enormous pressure to reach the deadline,' says Richard Hopkins.

You need to be resilient and resourceful in the face of a constant stream of dilemmas and conundrums. 'Something is going to go wrong, you're going to lose a location or the talent's going to pull out. It's always fire fighting in TV,' says Richard Drew. 'You need to be someone who's going to be positive and expect that things are going to go a little bit wrong. You need to find a way of overcoming those problems and solving those challenges and not be someone who's going to lose their head or get annoyed or moody.'

You need to have a contingency plan and a cool head. You need to be committed in order to immerse yourself in your production to the exclusion of everything else. 'You've got to have an instinct for getting on with people, for knowing what makes a good story, for working really hard and not getting too pushy. You've got to be dedicated to hard work and absolutely love telly,' says Liz Mills.

It's gruellingly tiring as most jobs are stressful, demanding and all consuming. 'Quite often it is long hours,' says Richard Hopkins. When in production, you have to be prepared to work weekends, late nights and early mornings (even through the night if you are working on a daily or live show) and do what it takes to deliver the programme. You might not see your friends and family until the show is over because even if you do have the time you're too exhausted to go out.

'You have to put your personal and social life on the backburner,' says Richard Drew. 'When I was show running I had to give up concert tickets, missed birthdays, all kinds of things. You shouldn't get yourself exploited, you shouldn't have to work 24/7, but you need to be able to put in the time when needed to get the show done.'

'When you're in your 20s you are prepared to undergo a lot of discomfort, big demands, no social life, to give in to your work,' says Moray Coulter. 'By the time you're 30 you're wanting a bit of your own life back, maybe you're having a family, maybe you're seeing more of your friends, maybe you want a hobby. People drop out as it's so unforgiving. You have to give your all.'

TV is a young person's game or someone with boundless energy, commitment and enthusiasm. TV can be exploitative and some of the jobs you'll do as a runner are menial when you start. 'You have to do a lot of rubbish when you're a runner,' says Kate Phillips of BBC Entertainment. You're on the lowest rung of the ladder, a chauffeur, gofer, getter of lunch and in some cases barely on the minimum wage. The pay is low, though you can progress very quickly if you're good and get noticed. 'The worst thing about telly when you start is the hours, but if you're good you'll move up fast.'

Dealing with the talent

As well as off screen talent there's the on screen talent – presenters, celebrities, actors, reality show contestants and contributors. They're essential to your programme but often they can be difficult, unco-operative, prone to tantrums and a nightmare to work with. Often they'll have their own agenda and ideas which may not be the same as yours; you have to manage that difficult relationship.

I've seen presenters square up to producers, swear at them and threaten to walk out just before a live show, refuse to ask questions, attend interviews and create havoc because they refuse to wear something. I've seen presenters reduce producers to tears, have them fired or vilified. They can seem incredibly confident and funny on screen but some may just repeat what they hear on talk-back verbatim. As the producer you have to be on the ball knowing all the questions, the running order, the content and the timing of the interview. You'll be juggling the team, the schedule, the budget as well as the talent and content. So you need to be focused and strong.

I've had to:
- Brief a notorious sex pest presenter and then ignore his advances when he 'accidently' brushed against my breast
- Silently take a tirade of abuse from a furious celebrity after a colleague failed to book them a cab

- Ask a shamed boxer who had been knocked out within 59 seconds of entering a ring to punch his way out of a giant paper bag
- Find hilarious contributors with extreme and usual hobbies/pets by phone and put them on a live studio breakfast show without meeting them first
- Find and vet (on the phone), test and produce a live five-minute item with new inventions and insane and eccentric inventors
- Persuade a pregnant mother of four disabled sons to come onto a daytime relationship show to discuss whether she would have an abortion if she was carrying another son with disabilities
- Find and then persuade the bereaved owner of a stuffed dead dog to drive down from Leeds to London overnight to appear on a show the next morning to celebrate their pet's life by performing a live tribute funeral
- Within a year of being in TV step in to direct a shoot I had set up when the director was fired and then edit the films I had shot for that week's transmission – with no experience of either
- Persuade an angry celebrity whose fee and travel expenses had been reduced to those his agent had agreed to go on air
- Calmly brief a late celebrity during make up, while the show was being transmitted live, to agree to and understand five different items before delivering them on air
- Find a double decker bus for filming an interview for the next day for free (no easy task when no bus company will release a vehicle from service – let alone for free)
- Direct and produce an item with a group of grumpy, naked, smelly and unco-operative poets without showing any dangly bits
- Found my job description had changed overnight and had to fight the world's press at film premieres to get access to and interview celebrities for next day's transmission (with no training or experience)
- Find factually accurate stories, with credible witnesses, which had the potential for dramatic reconstruction that could work as

12-minute films for a paranormal series, which hadn't been covered in the four previous series

- Answer the unreasonable demands of a bullying executive producer when I was in hospital with my mother who was having chemotherapy
- Create, produce, script, format and find talent for a 13-part travel gameshow in six weeks before six weeks of filming started with two teams filming around the world
- Devise, shoot and create the format for a 10-part gameshow with 200 films in six weeks then work for three weeks from 6am until midnight recording the studio shows
- Format, create and set up a 90-minute Saturday night gameshow pilot with 100 contestants filming around the country with a 20-foot egg in six weeks!

And me – how did I get my first break in TV?

I read history and English at the University of Sussex at a time when media studies degrees were pretty rare. It never occurred to me to train for or seek a job in TV. It seemed such an impossible task with no easily discernable career path and just six places a year on the BBC researcher trainee scheme to which thousands applied but for which only people from Oxbridge seemed to be accepted. I didn't have the confidence then to even try. I fell into television by accident.

I left university and worked as a journalist on a trade magazine before doing a post graduate qualification in journalism and freelancing for the music press. I believe that the skills I learned as a journalist were invaluable if not essential for a job as a TV researcher, as you use the same investigative and writing skills when putting together a brief, script, writing a list of questions, doing interviews or looking for new ideas and stories.

I got my first break in TV through word of mouth. I applied for a job as a music researcher on an entertainment show called *Big City*, a topical entertainment show made by Wall To Wall Television for

Carlton/ITV. Anna Richardson, a colleague from my journalism post grad course, told me that they were looking for a music specialist. She told me what she had been asked in the interview. I thought, 'I know that! I could do that!' and after several interviews with Nicola Gooch, the series producer and the executive producer, I was given a junior researcher position with a six week probationary period. The fact that I was a music specialist enabled me to make a shortcut into TV production, bypassing the usual first TV entry job – the runner. I was unusual in this respect but by no means unique. It is possible to go straight in as a researcher if you have a unique specialism and transferable skills to offer.

I was never given a clear brief or job description and I had no idea what I should be doing but I figured it out quite quickly. Using my experience as a journalist I found stories and came up with ideas, found and booked contributors, bands and celebrities, researched subjects and wrote interviews, briefs and scripts. I also found unusual (and free) locations, had to recce, set up and go on all the shoots – doing everything from writing the call sheet and arranging travel, parking, and lunch to finding any props, and supporting the edit by obtaining and clearing stills, music and clips.

These were all things I learnt as I went along, delivering what was asked of me – sometimes when I didn't have a clue where to look and didn't dare ask for fear of looking stupid!

It was fun and exciting but also frightening because it was constantly challenging. Although I was organised and planned ahead, I was always being asked for things at the last minute (and for free). Also, celebrities, bands, locations, contributors and stories could fall down at the last minute – as they always do. The schedule couldn't change with the show on air every week, so I'd have to find replacements at the last minute and rewrite and plan everything to fill our shoot day and the show.

It was stimulating but stressful. Failure was not an option. All the other researchers were all more experienced and more confident than I. I felt that I could be asked to leave at any time, but I worked hard and was promoted to researcher after my six week probationary period and worked on the show for eight months until it finished its run. I never lost this fear of failure. It gave me an unrelenting desire to deliver and achieve results and so I was employed continuously.

Before this first job came to an end I heard by word of mouth of a new Saturday night ITV 1 entertainment show. Planet 24 had a terrible reputation as the TV equivalent of a sweat shop – the hours were long and gruelling and it was based in the Docklands, miles away from anywhere and difficult to get to. Former employees warned me against applying but my boss, Nicola Gooch, said that it was the best training ground in TV I could wish for. I'd make contacts, learn new skills and gain invaluable experience; that if I could survive there I could work anywhere.

So I decided to apply. I had a challenging interview with the series editor who sat with his feet on the desk, watching TV and typing at the same time as firing questions at me about what suggestions I had for guests, VTs, studio items and new ideas. I was offered the job the same day. I started the week after finishing at Wall To Wall Television.

Where have I worked?

I've worked in production making shows for all broadcasters, programme development and recruitment. I've worked on live studio shows as live pre-recorded shows, on location and across entertainment and factual entertainment genres.

I worked for the TV production company Planet 24 (a now defunct indie but pioneers and creators of defining and groundbreaking TV programmes *The Word, The Big Breakfast* and *Survivor*) between 1997 and 2002. I had five different roles on *The Big Breakfast* from producing and directing celebrity VT packages and producing the live show to features editor.

As well as working on *The Big Breakfast* in several different roles I have worked as a researcher, assistant producer, producer and series producer. I've series produced six different entertainment shows for

Endemol, BBC Entertainment, RDF Media and Prospect Pictures amongst others.

For the last five years I've worked in programme development for RDF Media, Fox and Celador Productions and my credits include *Rock School* (Channel 4), *Brand New You* (five), *Who's Had What Done* (ITV), *The Tim Lovejoy Show* (Sky One), *Turn Back Your Body Clock* (Channel 4) and *World of Compulsive Hoarders* (Channel 4).

I've also worked in programme recruitment which is where the inspiration for this book came from. In 2005 I became head of talent at Zig Zag Productions, hiring and spotting creative talent, nurturing and promoting runners, mentoring the trainee on the Channel 4 researcher scheme and helping freelancers find work when their contracts with the company ended.

Since 2007 I've worked on a freelance basis crewing up shows and finding creative talent up to senior level for the BBC, Objective Productions, Impossible Pictures and North One amongst others. This role involves finding, matching and nurturing key talent to specific programme briefs and giving freelancers help and advice to help them achieve their potential.

So, if you're still with me and you think you've got what it takes read on because I'll show you how you start . . .

PROFILE

Andrew O'Connor, chief executive officer, Objective Productions
Andrew has had a varied career in theatre, in television as an actor, writer, producer and director, and executive producer. In 1997 he retired from performing to concentrate on running his television production company, Objective Productions.

As a performer, Andrew began his career as a child actor in BBC TV

series, *Canal Children*. Later he appeared regularly as a comedian and presenter in TV programmes such as *Copycats, Live from the Palladium* and *The Alphabet Game*. Andrew also toured extensively as a live performer, first as a magician and then as a stand up comedian. As an actor he played lead roles in the musicals *Barnum, Me and My Girl* and *Billy*.

As a writer/producer he has created/co-created over fifty TV series including: *The Quick Trick Show*, Derren Brown's TV series and specials, *The Real Hustle* and *Peep Show*.

Andrew's TV shows have been nominated for national and international awards and have won two BAFTAs, a Golden Rose, a Silver Rose, a South Bank Show Award, three RTS Awards and four British Comedy Awards.

Programme credits

2007–at the time of writing (spring '09)
Comedy Sketchbook Series (BBC), *Derren Brown Mind Control* (USA) (Sci Fi Channel), *Mike Strutter 2* (MTV), *The Real Hustle* series 3, 4 (BBC), *Return of 'Allo,' Allo* (BBC), *Star Stories* series 2, 3 (Channel 4), *The Peter Serafinowicz Show* (BBC), *Convention Crashers* (C4), *Derren Brown: The System* (C4), *Derren Brown Trick or Treat* series 2 (C4), *Comedy Live Presents* (C4), *The People Watchers* (BBC), *The Real Hustle Las Vegas* series 5 (BBC3), *Peep Show* series 5 (C4), *You've Got The Answer* (BBC pilot), *Kevin Bishop* series 1 (C4).

2006 Credits on co-productions
Indestructibles (BBC), *50 Greatest Television Dramas* (C4), *The Real Blue Nuns* (C4), *What the Pythons Did Next* (C4), *The Fame List* (C4).

2003–2006 executive producer
Derren Brown Russian Roulette (C4), *Greatest Magic Tricks in the Universe . . . Ever* (five), *Magick* (Channel 4), *Psychic Secrets Revealed* (five), *Secrets of Magic* series 1, 2 (BBC), *Comedy Heroes* 1, 2 (five), *Derren Brown Trick of The Mind* series 1, 2, 3 (C4), *4 Go Dating,* also co-creator (C4), *50 Worst Decisions* (Sky), *Best Unseen Ads* (Sky),

Britain's Favourite Comedian (five), *Derren Brown Messiah* (C4), *Derren Brown Russian Roulette Reloaded* (C4), *Derren Brown Séance* (C4), *Greatest TV Moments* series 1, 2 (five), *John Lydon Goes Ape* (five), *John Lydon Shark Attack* (five), *Monkey Magic* series 2 (five), *Peep Show* series 2, 3, 4 (Channel 4), *Secret World of Magic* (Sky), *Thomas Solomon, Escape Artist* (C4), *Britain's Most Watched TV* (five), *Undercover Magic* (Shock Magic) (Sky), *50 Questions of Political Incorrectness* (Sky), *Balls of Steel* series 1, 2 (C4), *Derren Brown: The Gathering* (C4), *Derren Brown: The Heist* (C4), *Dial A Mum* (ITV), *Dirty Tricks* (C4), *Extreme Family Values* (Sky), *Ghosthunter* (Sky), *Greatest Tabloid Headlines* (Scoop) (C4), *Massive Balls of Steel* (E4), *Miracle of Jesus* (C4), *Rajamn, The Evil Hypnotist* (E4), *Return of . . . The Goodies* (BBC), *Rewriting History* (C4 Learning), *Seven Stupid Escapes* (E4), *Top 50 Celebrity Animals* (Sky), *When Magic Tricks Go Wrong* (C4), *Ricky Meets . . . Larry David* (C4), *Deathwish Live* (C4), *The Last Word* (More 4), *The Real Hustle* series 1, 2 (BBC 3), *Tricks From the Bible* (C4), *TV Heaven, Telly Hell* series 1, 2 (C4), *30 Greatest Political Comedies* (More 4), *Age Concern: Celebrities Go To Seed* (five), *Derren Brown: Something Wicked This Way Comes* (C4), *Dutch Elm* (Paramount), *Fear of . . . Flying* (C4), *Greatest TV Cock Ups* (five), *Is Benny Hill Funny?* (C4), *Mike Strutter* series 1 (MTV), *Point Break: Celebrities Off The Rails* (five), *The Real Hustle Christmas Special* (BBC3), *Ricky Gervais Meets . . . Garry Shandling* (C4), *Ricky Gervais Meets . . . Christopher Guest* (C4), *Size Matters: Celebrities on the Scales* (five), *Star Stories* series 1 (C4), *Starkey's Last Word* (More 4), *Bullsh*t Detective* (BBC), *Derren Brown Trick or Treat* series 1 (C4), *Eejits* (C4), *Dog Face* (C4), *Perfect Night In* (C4), *The Real Hustle USA* (Court TV), *The Real Hustle: Winter Special* (BBC), *Twisted Tricks* (C4).

2003 – Executive producer and co-creator
Peep Show series 1 (C4), *19 Keys* (five), *Bedsitcom* (C4)

1999 – 2003 Executive producer at Objective
Producer: *Hyp The Streets with Paul McKenna* (C4), *Hypnosex* (E4), *Psychic!* (five), *Surprising Stars* (ITV), *The Quick Trick Show* series 3,

4, 5 (ITV), *5 Go Dating* series 1, 2 (C4), *Derren Brown Mind Control* series 2, 3 (C4), *Extreme Magic Extreme Danger* (ITV), *Meet the Challenge* (BBC), *50 Greatest Magic Tricks* (C4), *Monkey Magic* series 1 (five), *Movie Mistakes Uncovered Uncut* (five), *The Hidden Camera Show* (BBC), *Celebrity 5 Go Dating* (five).

1996–1999 Executive producer/Co-creator
The Quick Trick Show series 1 (ITV), *The Alphabet Game* (BBC) also presenter, *The Quick Trick Show* series 2 (ITV), *Derren Brown Mind Control* 1 (C4).

1996 – Founded Objective Productions

What does your job involve on a day-to-day basis?
I format and co-executive produce shows and support our executive producers. All five heads of department and the executive producers report to me. I pitch ideas two or three times a week. I help run the business on the bottom line trying to grow it. Since the company sale I have to report upwards on how it's going and what's happening.

Your output is incredibly prolific and your hit rate amazing – how do you do it?
Objective has created a brand and environment where talented people want to come and work and then, crucially, stay. There hasn't been a single producer who I rated and I really wanted to keep who has ever left the company. It's all down to the incredibly high quality of the ideas and the execution of those ideas of the people who work here.

What inspires your ideas?
God knows!

How did you get into television?
I was a performer up until 1997. Before that I had written some kids' dramas and some game shows. I had some producer credits in my late 20s/early 30s. I hosted some shows that I wrote, like the pilots of *Raise The Roof* and *Lose A Million*. I wrote the *Quick Trick*

Show with a producer who ran his own production company and I had a deal that when they sold it as a format I would get half the format fee, half the executive producer fee and part of the production fee. After he sold his company he couldn't afford to give me the same deal so I formed my own company, Objective Productions, and started pitching. The first thing we did was *The Alphabet Game* which was a daytime game show for the BBC. I hosted it. The first series was made from the front room of my house with a producer, two researchers and a production manager!

We had quite a lot of success with that, it kept coming back. Then there was a Saturday night show called *The Hidden Camera Show* with Ainsley Harriet. So we proved quite quickly what our brand would be – comedy entertainment. The first few years were great. We had a daytime show called *Meet the Challenge* for BBC 1, but then that didn't work and then *Hidden Camera* didn't work and we had a tough couple of years.

Did you have a game plan?
The long-term plan was to have a company that I cared about and loved.

What advice would you give to someone starting out?
Make shows that you want to watch. That's the key thing. The shows that I make are ones I would watch if I wasn't at Objective. Decide what you like to watch and then find out which companies make those shows. You should work for free at those companies until you're fifty! Your job is to go to a production company and make yourself indispensible so they're unable to cope without you. Anthony Owen came here as a magic researcher and now he's head of magic.

What skills have contributed to your success?
Being able to create, sell and make an idea. From that everything will follow. Setting up a production company is a simple thing. You can get people to help you run it.

What's the best piece of advice you were given?
Never give away a bit of your company to someone who is not
creative. You can always find people to do the business side of it –
balance the books, sort out the HR – but the most valuable thing
in British television is the people who can create, sell and make
successful programmes. If you can do one of those three you can
have a career but if you can do all three you can make a lot of
money.

What have been your best moments working in TV so far?
The success of *Peep Show* and recording *Russian Roulette* with
Derren Brown in Jersey. It was an amazing feeling when it was
going out live. What I really care about is making shows that
engage the audience, that's the best bit. But I love the process of
making TV even more than that – having an idea, working on it,
getting into the edits, working out if it any good, crafting it. You've
got to have fun during work.

What's the worst thing about working in TV?
Nearly going under wasn't very good. I was cashing in pensions
to stay afloat. Things were so hard at that stage. And the
competitiveness. I still get upset when a competitor has an
idea that I wish we had made.

What personal qualities have helped you?
Tenacity and whatever creative judgement I may have. I've learned
that from the outside other people's careers may seem blessed and
easy but from the inside it is a great struggle. I feel like my career
has been a huge battle.

What's the best way to get a job in television?
Find a show that is already successful and formatted that you
love and do whatever you can to get a job on that show – offer to
make the tea!

Chapter Two
What you need to get started in TV

Love TV?

Do you love watching telly? Are you mad about it? People who work in TV are ambitious, single minded and committed to making programmes. To succeed in it, you must love TV to the exclusion of everything else. 'Above and beyond anything, you have to love the medium you are putting yourself into and you need to watch a hell of a lot of it,' says Grant Mansfield, managing director of RDF Television.

If you want to work in the industry you have to be passionate about programmes, love watching telly and be knowledgeable and critical in your viewing. You need to know what you like and what don't like and why.

'You have to be not just interested in but passionate about television. It's an incredibly competitive, all consuming industry and you're not going to get far unless you're really dedicated,' says Matt Born managing director of agency DV Talent, which represents top end shooting producer/directors and APs. 'It sounds obvious but it's remarkable how many people write to us for advice on getting into the industry but don't actually watch anything. They seem to want a career in television simply because it has some an association with celebrity.'

If you love TV it's easy and enjoyable to watch lots of it and for it not to really seem like a job at all. 'I knew I wanted to do telly since I was ten,' says researcher John Adams who blagged his way into the studios and gallery of *Noel's House Party* when he was still at school. 'I loved TV,' says producer/director Andrew Chaplin. None of his family worked in the media but Andrew got work experience on a lottery show at the BBC when he was 16.

'I always wanted to work in TV because I spent the whole of my adolescence plonked in front it,' says series producer Alex Marengo. 'It's about deciding you want to make programmes.'

TV is tough and competitive. Above all it's a serious business about making money, and with budget cuts, one that demands long hours as just a few people do the work of several. It relies a lot on unpaid or badly paid labour at the lower end of the scale. In short, it's bloody hard work.

If you just fancy the idea of TV and have only a vague interest then forget it. You have to be desperate to work in the industry and be prepared to do anything to get started because you will be competing against people who are equally determined. Eagerness, keenness and commitment go a long way when you are trying to get a foot in the door.

'When you start off be prepared to do anything, and be incredibly willing and interested. Enthusiasm will get you such a long way, further than you could possibly think,' says Daisy Goodwin, managing director of Silver River. 'Somebody who is really prepared to try hard, even if they get it wrong, the fact that they have tried goes down incredibly well.'

I interviewed candidates applying for the Channel 4 diversity researcher's scheme. I was aiming to give a complete newcomer the opportunity to come to Zig Zag for a year. It was an amazing chance to get a job as a researcher and to work on different productions and also receive skills training at Channel 4. It was a paid job which bypassed the typical runner's role.

One of the questions I asked was 'Why do you want to work in TV?' An alarmingly high number of the shortlisted candidates were vague and unsure, saying without confidence that they thought it might be interesting but also fancied working in PR or film. Wrong answer! Needless to say their applications didn't go any further. No matter how good they were, they were far too casual in their approach. They were competing against far more confident and passionate candidates brimming with ideas and enthusiasm.

'Friends ring me up and ask me to have a chat to their sons or daughters who want to get into TV,' says Alex Marengo. 'But when I speak to them they haven't got the foggiest idea. They don't watch TV or have any reason why they want to work in it or what it involves.'

TV is all about ratings, so you must have an awareness of what people are watching and why. All broadcasters are chasing ratings and audience share which in the age of digital free to air TV has become increasingly fragmented. Decisions about programming are based on what is hot, what's being talked about at work the next day. If you don't watch TV how can you be part of that debate or have anything to contribute? Employers look for this passion when they are hiring. You need to watch the hit shows.

'I look for someone who loves television,' says Kate Phillips of BBC Entertainment. 'It's amazing how many people I interview who don't watch TV or they come for a job in entertainment and they haven't watched *Strictly Come Dancing* or *Joseph* or *Saturday Night Takeaway*. You need to love television and want to work in it. I decided to work in television because I spent all my childhood watching it. My mum would say, "Go outside and play" but it's amazing – the knowledge I have is actually useful in my job now.'

How well do you know yourself?

Most creative people in TV production are self employed freelancers who move from job to job, short term contract to short term contract. They may change company and programme up to three times a year, maybe more; they face job insecurity and periods of unemployment. Freelancers are 'sole traders', small businesses responsible for pitching for work, invoicing when it's done, keeping proper records, paying their own national insurance contributions and putting aside a percentage of what they earn (at least a third) to pay their tax. Are you the kind of sensible but creative person who can do this?

'You have to be incredibly organised and understand it's a free-lance industry. You have to save your own tax and you should expect to be unemployed so you should put money aside for that too,' says

AP Claire Richards. It's OK when you have no dependents, financial commitments or personal interests, but it's still very insecure which makes it difficult to plan for the future.

'You have to be prepared to be poor and struggling when all your peers are in stable, well-paid jobs,' says producer/director Anna Keel. You need to work out whether you are suitable for this style of working and whether it is something you will enjoy.

'A lot of my friends have already left the industry; they've given it up and gone to do other things like teaching as TV is so insecure and you can't get a mortgage,' adds Claire, 'but I've never been unemployed in my three years of working in TV, I love it and don't want to do anything else.'

You also need to decide whether working in production matches your interests and personality. 'You need to look at how you present yourself,' says Julia Waring of RDF. 'Are you the kind of person who can go to shopping centres and get contributors? Or are you the kind of person who is best at fact gathering or investigations? You need to know yourself.'

There are several different genres of TV, from studio-based entertainment or game shows and reality series like *Big Brother*, right through to serious, investigative documentaries and high end specialist factual series. You need to work out what genuinely interests you, what you're most suited to and which would be most appealing.

'I interviewed one girl recently and she was really apologetic about her favourite programme, *Big Brother*,' says Kate Phillips. 'You should just say what you love and not be worried about it.'

What do you really like watching? What would you enjoy making? Does the excitement of a live studio entertainment show appeal to you? Or would you enjoy immersing yourself in telling an amazing story in a factual documentary? You shouldn't pick a genre if it doesn't suit your true personality. Be true to yourself and keep it real. Don't feel that you have to make an observational documentary because it seems more worthy. Choose something that you would genuinely enjoy making.

To decide what genre would suit you work out what you watch and like. 'People should be realistic about their own ability and should think about the genre of television they are genuinely interested in,' says Julia Waring. 'You mustn't lie about what you are interested in or you won't be able to give your heart to that particular type of programme.'

Try to be as objective as you can to assess your positive qualities: working through them with someone else is a useful exercise. 'I don't think you can do it yourself,' says Moray Coulter of ITV Productions. 'You need to ask somebody else to be a resource for you. It's a good idea to speak to a friend who knows you well and to sit down and work through what your weaknesses are from their perception of you. And if you agree, work out what your strengths are and how to make the most of them.'

It's an exercise that works well at any level and Moray uses it himself. 'It goes right up the scale. You're never too old to do it. I did exactly the same thing recently. I had lunch with a friend and we worked out what sort of things I am good at. It was very useful. I had just left one job and was trying to work out what I should do next and what I could offer.'

What qualities do you need to work in TV?

Most people working in TV share common qualities. These are:
- Stamina and a capacity for hard work
- Managing well under pressure
- Working well in a team and being a good team player
- Being a self-starter able to work on your own and use your initiative
- Being easy going, friendly and pleasant to work with
- Being confident and tenacious
- Having the ability to charm, flatter and blag!

TV is generally a young industry because it requires enthusiasm and stamina. 'You've got to be hard working, energetic, optimistic and reliable,' says Daisy Goodwin, managing director of Silver River. 'You need to be able to get on with people, be persuasive and charming. If you're not charming, you have to be incredibly talented.'

Being prepared to work hard is essential. 'I'd always hire a hard worker above someone who is talented,' says Richard Drew, a successful British executive producer now based in America. 'Someone who is going to put in the hours and really work their butt off. The most important quality you should have is a hard work ethic.'

You have to be happy to sacrifice your personal life for your job and accept that your job will take over your life, regardless of any plans. 'You've got to be ready to work 24/7 to make the show work if needs be. If the night before your show you lose some key talent, you have to be willing to stay until the early hours of the morning to get them back or find a replacement. You have to be prepared to work all weekend or whatever you have to do to make that show happen,' says Drew.

'The qualities you need to succeed in TV are commonsense, good people skills, calmness under pressure, imagination, a work ethic,' says Conrad Green, the British executive producer of *Dancing With The Stars* in the US.

And crucially, never whinge when you're finding it tough – just get on with the job. He adds. 'Never moan about how tough it is for you – absorbing problems and dealing with them will get you promoted quicker than anything else.'

As a series producer I really valued the hard working, resourceful members of my teams who resolved problems and gave me updates with their proposed solutions rather than constantly bringing problems to me. It's preferable to work with them rather than others who are a drain on your time, asking you to solve their problems and do their work for them. Being able to work things out for yourself with little help from anyone else benefits the production and yourself enormously.

Working long hours is tiring and it is tempting to moan and pass the buck as exhaustion takes its toll. You need to be resilient and have plenty of stamina to get through a busy schedule, especially when you're feeling under the weather. As the workforce is freelance and people are scheduled to work certain days there simply isn't time to be ill. Daisy Goodwin takes a dim view of people taking time off sick. 'I'm pretty tough about being ill,' she says. 'I think that if you really enjoy your job, you're never sick. There's an attitude to work that I really notice when people are here. If I saw someone who had taken four days off a month, I would wonder what's that all about.'

And she's not alone, being ill can let down the rest of your team who have to cover for you.

How you manage stress, how strong you are mentally and physically will have a bearing on how well you can do your job. If you can cope with long hours, thrive on stress and problem solving then you'll enjoy working in TV. Claire Richards finds she can easily deal with the pressure and the long hours because it gives her a buzz. 'When I'm out filming I get an adrenaline rush. I only get tired at the end of the day.'

'You've got be able to work well under pressure,' says Richard Hopkins of Fever Media. 'Pressure is always on to reach the deadline. That's a scary feeling and, to be able to manage that fear, it's about organisation really, being as anal as you can about how you line up all the things you need on the day, and how you work with the people around you, above or below. People above you won't always be brilliantly talented, you won't always work with the same team, you might suddenly get a job on a show where your series producer or producer is not very good and you almost have to make up for their weaknesses if you are an AP or researcher.'

Working well in a team and getting on with people when you're pushed to breaking point and are juggling a million demands is crucial. Being calm under pressure and gracious in the face of enormous adversity goes a long way.

'I've always thought I'm friendly and approachable and I never lose my temper,' says assistant producer Jenny Popplewell. 'It pays to be nice. It helps to make an enjoyable work environment. And it also helps you build working relationships with contributors, especially as sometimes you need to ask a lot of them.'

It's important to be able to work well as part of a team, to get along with your team members and be collaborative. 'On a lot of big shows, like the *Paul O'Grady Show*, you're working so hard that you completely forget about the rest of your life,' says researcher John Adams. 'You've made this new family because that's been your whole life, so it's good if you get on with those people.

The cliché 'there's no I in team' really does count for a lot. Series producers are responsible for hiring people for their productions and putting teams together. 'The most essential thing is to be someone who is used to working in a team. Some people thrive on conflict. I don't. Making TV shows is a collaborative process. Putting together a nice team of talented people who you like and trust is a huge benefit to the whole production,' says series producer Alex Marengo.

And to your benefit too; if your fellow team members become your friends they'll help you by telling you about jobs they hear about and recommend you for work. Jenny Popplewell has worked in television since 2004. She says, 'I've never applied for work. I've always heard of jobs through word of mouth. Friends I've worked with go on to other productions with different companies and they call me up to tell me about jobs or suggest me to other people or hire me themselves. I've never put my CV on a website or had to look for a job and I've been employed continuously.'

Never be rude to people on the way up – or in fact ever! As it's a freelance marketplace you'll come across people again and again and the state of these relationships will help or wreck your career. 'Be your best,' says Claire Richards. 'Never lose your temper or be rude. There's no point in that. You may need to ring them up at some point in the future.'

As it's a word of mouth industry word will get around of how popular you are or if you're not. 'References can be difficult because you can get on with someone really well and another not so well,' says Michelle Matherson, talent executive at BBC Factual. 'You tend to be as good as your last job, so I look at two or three references if I get a bad feedback from one. Think carefully about who you put down as your referee because they will be checked. You can have a glittering career and then come across someone who doesn't like you.' And you only need to get a couple of bad references and your career can stall.

'Bad references travel,' says Matt Born of DV Talent, 'which is incredibly frustrating if it's the result of a personality clash rather than anything more concrete. So, if you want to be well thought of it's vital to get on with the different teams you're working with. That's not always easy in telly where there are a lot of strong personalities, and it doesn't mean just flattering the boss; success is often about winning round the difficult characters on the production team as much as the difficult contributors.'

Although you should be single minded in your determination to succeed, it shouldn't be at any cost. You will come across the same people again and again – sometimes years later, but memories last a long time. You can't be rude and ruthless and not expect it to come back and haunt you. 'Being a nice person to work with is important,' says Richard Hopkins. 'Nobody wants to work with somebody unpleasant. Even though they might go a long way sometime or other in the future, the wheel will turn full circle.'

Bullying is endemic in TV and it is sometimes tolerated for longer than it should be, but bullies do get their comeuppance in the end. One executive producer found he was blacklisted when the show came off air. When he was thrust out into the market place he couldn't get work. All the runners, researchers and producers he had bullied and shouted at now worked at different production companies and spread the news of his bad reputation; he found it difficult to find work despite his experience.

Similarly, a successful commissioning editor for a terrestrial broadcaster found it impossible to find work when she left the channel as she was also an unpleasant bully who had abused her power. She had a terrible reputation and was so difficult that some production companies even refused to have her oversee their programmes.

When she tried to get work in production she found no one wanted to work with her or hire her.

Career progression is fast in TV and it's not uncommon for people to change roles and for power to change hands. 'Be nice to young people,' says Lucy Reese. 'Some of them will end up being your boss!'

The BBC's Kate Phillips says, 'Today's runner is tomorrow's channel controller is a saying that's really true in telly. It's swings and roundabouts. One minute a person who used to be your researcher becomes a commissioning editor and next year you're actually pitching to them!'

As well as getting on with people off screen, you'll need to be able to win the trust and confidence of the on-screen talent. 'You've got to be organised and be professional about how you look for talent and how you face contributors,' says Richard Drew.

If you find talent or work with them at the beginning of their career and become friends with them as their star ascends, so will yours. As they become famous if they rate you they will want to take you with them as they may insist on working with a producer they trust. Many celebrities will have a 'chosen' producer or only work with certain producers – and refuse to work with others! If you develop a solid and true friendship with someone they will want to work with you again and again, and this will ensure continuous work – as long as they are popular and working themselves!

Being confident is essential for working and surviving in TV. It's important to believe in yourself and have the innate ability to ask questions of strangers, to mix, talk and engage with different groups of people. Being a 'people person', using your intuition to get along with others and win their trust, and having common sense and to be able to think are key skills.

'If you panic or you come across as lacking in confidence people won't have faith in your ability to do a job. You have to have the confidence to ask for help or information,' says Claire Richards.

Inspiring confidence is also important, being seen to be able to do all that is asked of you and more and delivering. Always say yes to a question and find out the answer later. And if you don't know, find someone who does. Quickly!

'The most important thing to have when trying to get into this industry is confidence,' says Jo Taylor, talent manager at 4Talent. But don't be over confident. 'Sometimes that confidence can manifest itself as arrogance, like someone saying that they are fantastic but they don't have the experience to live up to that. You don't have to be really, really pushy to get ahead.'

But you do need to be able to blag, flatter and charm to win people over and make them like you. To survive in TV you have to be a consummately professional persuader who's likeable, popular and can handle other people and yourself in difficult situations. Never underestimate the power of flattery but be sincere and don't over egg it. Be genuine in your approach or people will see through you.

Claire Richards has won over difficult contributors by having good manners. 'It's about getting on with someone. Be approachable, be nice and sound like you're going to do the best job. If someone asked me for help I would give it to them if they were extremely nice and polite and if they flattered me. I would give an hour of my time. It's about being yourself and nice. Don't be false; people will see through it.'

'I am courteous but tenacious,' says Simon Warrington. 'This allows me to speak to various levels of people on a common ground that helps me achieve what I'm looking for.'

What qualifications do you need to work in TV?
The number of graduates entering the industry is rising and according to the skills sector council for the creative industries in Britain, Skillset, people working in TV are highly qualified with two-thirds having a degree. But it is possible to succeed in TV without going to university. 'You don't necessarily have to have a fantastic degree but you do need to be intelligent,' says Daisy Goodwin.

There are people in the industry who haven't been to university and some who don't have any qualifications at all. Andrew

O'Connor, the chief executive officer and founder of Objective Productions, doesn't have any formal qualifications. 'At 14 I came to London and went to drama school,' he says. 'I was the youngest student in the year.' He then became an actor and then presenter before writing and executive producing TV shows. Andrew doesn't have a degree and doesn't think it essential to work in television. 'I don't think it matters to be honest,' he says.

Director Andrew Chaplin did A levels but gave up a place at university to take a job on T4 as a runner. 'I passionately believe, one hundred per cent, that you don't need to go to university,' he says. 'Not having a degree hasn't held me back. My friends would ring me up at three in the morning from university saying they were drunk and having a great time. I had to get up at six in the morning and was working seven days a week but I have no regrets. It was hell but I loved it.'

It's also possible to do other things and then come to TV later. Assistant producer Simon Warrington was a professional footballer for eight years, playing from the age of 14 as an England schoolboy and then for Bradford City Football Club before having to retire due to an injury. He has no formal qualifications but worked in graphic design and the music industry before getting into TV in 2003 and has successfully worked in the industry ever since.

You can enter the industry at a young age and start at the bottom. Suzy Lamb is a successful entertainment producer now working in the United States. She started at the BBC as a PA when she was just 17 years old and worked her way through the ranks of the Corporation for over 20 years becoming a long-serving member of staff and gaining an unrivalled reputation as a producer of live entertainment shows. She series produced all the live lottery shows from *Jet Set, In It To Win It* to series producing *How Do You Solve A Problem Like Maria?* and *Dancing With The Stars,* the US version of *Strictly Come Dancing.*

'I don't think there are any actual qualifications you need if you have the skills, desire and tenacity to work through the early years in the industry. Talent will get spotted and be rewarded; it's a very meritocratic industry in that respect,' says Conrad Green, the multi-award winning British executive producer of *Dancing With The Stars* in America.

'If you look back 20 years, when fewer people went to university, there were people running television channels who didn't have a degree,' says Julia Waring of RDF. 'Nigel Pickard, founding controller of CBeebies and CBBC, now chairman of Foundation, part of the RDF Media Group, isn't a university graduate. I think it is something that has evolved in the last 10 years.'

'I don't think that qualifications make any difference at all. Obviously it's useful if you have got good academic qualifications, depending on the type of programme you're wanting to go into. I'd rather people who decide not to carry on with full time education after A levels try to get into the industry rather than spending £20,000 on a media course at a fifth rate university. It won't be worth the paper it's written on,' she says. 'There are a few really good universities which provide solid practical courses but those people would still have to come in as a runner.'

Waring accepts that a good degree on a CV will make a difference as to whether she will meet someone or not. 'I'm interested to meet people who, on paper, are quite clever because you want clever people but you can meet somebody who has graduated with a 2:1 from Bristol or Oxford or wherever, and they have absolutely no communication skills whatsoever. They are dead from the neck upwards. Then you meet someone who has left school or could have a 2:2 or no degree but they just have a charm about them. You know if you were a contributor that you would be soon under their spell and would do whatever they wanted you to do. And that's the key, isn't it?'

Jo Taylor of Channel 4 believes it is difficult to get on in TV without a degree. 'I'd really like to say that you don't but I really think you do. You need a degree in something, it doesn't need to be a degree in media, but I think it shows a commitment and it shows a level of learning that you don't get from coming out of school at 16. I don't think you have to have a degree from a fantastic university, you can have a BTEC or HNDs but having some more qualifications than just GCSEs really helps.' She adds, 'I didn't want to go to university; after my A levels I wanted to go and work in post production. I truthfully believe I wouldn't be where I am if I had done that. My husband came out of school at 16 and has struggled throughout his career because he hasn't got a degree. He's a production manager. It's taken him a long time to get to that level because people look at him and he's got only 9 O levels. They are good O levels but people say, "They are only O levels." They look at that. There's a prejudice there and he knows it.'

I know a very successful drama director who when he started out as an entertainment researcher added a fictitious degree to his CV. He never went to university but when he started looking for TV work he soon realised he would be overlooked without one on his CV and he still has it on there now – years later! Fortunately no one has ever checked.

'I think there are a lot of bright people who don't go to university and many not very intelligent people who do get on to university courses,' says Julia Waring. 'Who is the intelligent person? Within the communications industry, you need to be able to communicate.'

Do I need to do a media degree to get a job in TV?

A degree is useful when working in TV, but it doesn't have to be a media studies degree. 'I want to know that somebody is smart and curious and able to do research,' says Daisy Goodwin. 'And I'm not entirely sure that media studies degrees are the way forward. I don't think you need them. I would be much more interested in someone who studied an arts degree, or who had a good degree in any other discipline because I actually want someone who knows how to find things out, who's clever.'

'Media degrees are sometimes useful and sometimes misleading,' says Richard Hopkins. 'If you've got camera skills that's very useful because I know I can take you on as a runner, I can send you out on a shoot and you can hopefully make me a taster tape for a new show

I'm developing. It wouldn't necessarily mean you're going to be a good writer and it doesn't necessarily mean you're going to be a creative person. Quite often people with media degrees have disappointed me slightly,' he adds, 'because I've expected so much from them and it turns out they're not that great at shooting and the movie they've told me about in the interview isn't as brilliantly created as I thought it was going to be. It only really takes a one day course to teach someone how to use a Z1 or a DV, so it can be overestimated how important it is. But if you are really obsessed with television and want to go on and do a media degree, don't let it stop you. It will probably give you a slight advantage over a geography degree but I would say if you had a degree that involved a lot of writing and creativity it would probably put you on an equal footing.'

Tim Hincks, chief executive of Endemol UK says, 'If it's a degree that you enjoy doing then why not do it? I did my degree in politics because I thought I would enjoy it and I was interested in it. So, if you're interested in doing media studies then do it, but I don't think it gives you any more of a chance of being involved in television content than doing a medical degree.'

'People used to sneer at media courses,' says Lucy Reese, 'but I think a good media course can be helpful, especially one that has links to the industry. A media degree can be fun as well – which is more than can be said for many degrees!'

When Margaret Hodge, the former higher education minister, infamously branded some university courses as 'Mickey Mouse degrees' the comparison with the cartoon character became synonomous with media studies.

'There is a snobbery about media degrees,' says Michelle Matherson. 'But actually they are quite advanced and the modules within them take in all aspects of TV production, so I don't really feel we can be so snobby about people who have these types of degrees. It wouldn't put me off someone.'

It's just a case of careful research, being selective about the course you choose and deciding whether it is right for you. For example,

Skillset, the skills sector council for the creative industry, accredits a number of courses for TV in its Media Academy. This is a national network of colleges and universities that work with industry to develop the new wave of media talent. It is made up of 17 Media Academies, drawing together creative education partnerships from 43 colleges and universities across the UK. The institutions in the network are already centres of excellence in television.

'We went around the UK looking at universities and further education colleges, trying to identify the ones that really have industry-focused courses,' says Alice Dudley, training manager at Skillset. 'It's really hard for young people to find the right university course and do a media degree that will really help them work in the TV industry. I think if you actually want the practical skills to learn how to work in TV you need something a bit more focused.'

According to the Higher Education Statistics Agency (HESA), between 2001 and 2006 the number of students enrolled in media courses grew from 13,600 to 26,700, and the number of journalism degrees doubled. Because of the demand for media and film courses their numbers continue to grow. With such a variety of media studies courses on offer it's worth researching them thoroughly to find the ones that are more vocational and rather than academic.

'Not many media courses have a practical input from the TV industry,' says Alice Dudley. 'When people start off in TV they are going to have to be a runner and they need quite practical skills and I don't think a lot of those media degrees focus on those kinds of skills. They focus on other areas of knowledge.'

It's about being selective and researching them thoroughly as some offer a fantastic grounding. 'I did a three year media studies degree,' says executive producer Conrad Green, who gained a first class honours degree from the University of Westminster (formerly the Polytechnic of Central London). He says, 'I actually specialised in radio as I wanted to learn about story construction and how to build editorial, which I think was the right decision as college TV courses will always lag behind real-life production as a training ground.'

'When I was at uni I wrote and shot a documentary which won awards,' says Claire Richards who has a BA in Media Production from Staffordshire University. 'But I don't think my media degree helped me get into TV but it was useful technically; I could hire equipment and edit in my spare time.'

Research your course as you would a job or a project. Work out what you are genuinely interested in, what you want to know and find a course that matches your interests.

Find out:
- What's on the syllabus?
- What validation or accreditation do the courses have?
- Who are the lecturers teaching the course?
- How much industry experience do they have?
- What practical skills does the course have to offer?
- What equipment will you learn to use?
- How much hands-on contact will you get?
- How strong are the university's industry links?
- Do they offer work experience?

If you find the right course, you'll meet fellow students who could help you in the future as they progress through the industry, especially if you share the same skills and commitment.

Top TV Academy offers training to people wanting to get into or progress in TV. 'We get some students from media courses who don't know the basics,' says Liz Mills, founder. 'They won't know who's who in a production team. We have to fill in the gaps in their knowledge about what's on TV and going out at the moment. It's quite staggering really, their complete lack of knowledge of the very basics.'

'I don't think my media studies degree helped me find work,' says Claire Richards, who wishes that she had not done a degree but had started working in TV after A levels. 'I could have gone in sooner and started at the bottom and worked my way up. University helps you to grow up and find yourself and become independent but from a career level, I would have preferred to have three years more industry training on the job.'

This is a sentiment echoed by Julia Waring, who's responsible for hiring creative talent, overseeing the intern scheme and hiring runners at RDF. 'I feel very strongly that university courses like media studies need to be selective. It should be as rigorous as if you were applying to do law, where I assume you have to do some sort of test to see if you are able to do the work. Leeds University, for example, is very selective in their calls and generally the standard of people on their broadcasting course is very good; so is Bournemouth. But there are a lot of universities which really bump up their numbers by running media courses and somebody comes out in debt and they know nothing. They are utterly inappropriate and they think they are a producer/director but they just know nothing. It doesn't matter to me whether you have done engineering or media studies; it's solely about how you are as a person.'

Researcher David Minchin, who has been working TV since the beginning of 2006, agrees, 'Work experience people and runners who come out of university with a good media degree think they know everything but so many people I've worked with say they don't care what degree someone has. Real life is so different. It depends on how hard you work and how good you are.'

But having a good specialist degree in history, a science, anthropology or the arts can be helpful if you want to work in specialist factual programmes where your knowledge can give you a good starting point of entry. 'A degree is essential, especially if you're working on a specialist factual programme such as a science, history or archaeological programme. If you have to talk to academics and synthesize factual material it's a prerequisite,' says specialist factual series producer Alex Marengo.

'I find out whether someone has an understanding of the subject,' says BBC Talent executive Michelle Matherson, 'because this specialist knowledge could be useful to the programme.'

'It's a matter of where you can best slot yourself. If you have a fantastic physics degree, well then you are going to approach somebody who is science based,' says Julia Waring.

Studying languages at degree level is also incredibly useful, especially if you are working on travel programmes or series where you have to set up shoots, find contributors and film on location abroad.

'Being bilingual is a real asset,' says Julia Waring. 'One of my friends was intent on doing in media course and I advised her to do something else that she was really interested in. She chose Spanish and French. Fabulous! She has graduated from Leeds University with a Spanish and French degree. Her most logical course would be to go to any company that makes travel films and say, "I am a runner, I can drive, I can speak fluent French and Spanish". That's an immediate advantage over someone who doesn't speak any languages.'

David Minchin is fluent in three languages, including French, which gave him his first break as a researcher. 'I don't think it initially helped getting in the door,' he says, 'but I think that it definitely helped me get a job on *Wish You Were Here?* because there was a lot of travelling in Europe and as I speak Spanish and Italian it helped with that job, which was great.'

What skills do you need to work in TV?

Once people get in to TV they tend to move around, freelancing at different companies and being hired according to their reputation and the last credit they had on their CV. As the saying goes, 'You're only as good as your last show.' Reputation is everything and there are no hard and fast rules about how you gain a good one! Spotting who's making good programmes and having hits is always a good start: aim to work with them.

People who survive and succeed in telly tend to share a mixture of transferable skills including:
• Good creative skills – coming up with ideas for shows and working on them

- Good communication skills, both written and verbal
- The ability to sell yourself and your ideas to employers
- Being able to handle rejection.

Ideas, ideas, ideas

Ideas are a precious commodity in TV, a currency that increases your personal value. If you have a constant stream of good ideas and you can sell them, whether to a show or company, you can do very, very well indeed. If they can be made cheaply and easily, even better! Being creative, being able to write and come up with ideas for shows, games, contributors, presenters or titles for programmes from scratch is a key skill to working in TV, and can give you the edge for getting ahead.

Whether starting out or working as a jobbing freelancer, taking ideas to interviews is essential. 'I am always impressed when people come with ideas,' says Michelle Matherson. 'It's not just about the job that you're coming to be interviewed about. It's good to come and sell yourself because that way you stick in someone's mind.' It will also give you an advantage over your competitors and as a newcomer will give you the chance to grab the attention of your interviewer. 'Ideas are crucial in getting your foot in the door, which is always the toughest part,' says Grant Mansfield, managing director or RDF Television. 'If you assume that everybody who comes along for an interview is clever then having loads of ideas will give you the edge over everyone else. It gives you an instant head start.'

Alex Marengo has worked in TV for 17 years. 'I've got to the stage now where I go from one long project to another. People ring me up and offer me projects. Answering ads has never borne any fruit for me. It's a better strategy to work out what you want to do and approach those companies and have ideas to go in with. But never your best ones. Keep those in reserve! Put yourself about in a positive way; you get results if you go in with something to offer.'

Daisy Goodwin thinks it's unwise to be too precious about ideas. 'Don't be too queeny about something. People will say they've got a great idea but they can't tell you about it because you might steal it. A great idea is any great idea if it is being sold by a great person, so you have to be realistic. The most important thing you can do is find an environment where you're going to get breaks.' And to get that break you need to get your foot in the door and taking ideas can help you get that all important first break.

Your ideas don't have to be fully formatted, but they do need to be thought through and properly planned. You can only do this by spending time working them out in advance.

'If you're going for a job on a brand new show and you don't know the format, you should still be prepared,' says Richard Drew. 'I call ahead and let them know what the show is and talk them through it over the phone and I then ask them to prepare something. That's a really good indication of how badly someone wants the job. If they come across as very scrappy, that's not very impressive. I do care about the quality and how good the ideas are but also how much time they have spent on them, and how much they have thought them through.'

You could come up with new ideas for new programmes, or you could suggest ideas based on the company's existing style and brand, or you could come up with new ideas for existing programmes.

'You have to freshen up existing formats every time and push them further,' says Kate Phillips, head of development at BBC Entertainment. '*Big Brother* is probably the most successful format in the world, but every time you see a new series the audience expects more because they are getting more sophisticated. The first episode was so shocking because of the idea that everybody was being filmed 24 hours a day, on the loo, in the house but you get to series seven or eight and that's not shocking anymore. So, every time you have to push the boundaries further. The *Big Brother* team is really good. Not only are they having to produce so many hours of television but they have to keep coming up with the new ideas and games.'

Make sure you do your homework about the company and their shows. Don't suggest something that is already on air, or has been on air and done badly! Do turn up prepared with notes on funny titles, talent thoughts and ideas for shows that you would genuinely like to see and why you think your ideas would work.

'It's about a passion and having opinions,' says Jo Taylor. 'When you go to an interview you should have an idea of what programmes the company makes so when someone asks, "What programmes do we make?" you should be able to reel of a couple and say why you like them. You should have done your research about that company thoroughly before you step through the door because you may be the twentieth person who the talent manager has seen that day. What is going to set you apart from the rest? It's not just your personality, it's your passion, it's your enthusiasm, it's your communication, and also it's your knowledge.'

If you work in TV production on a particular show it's expected that you'll come up with ideas for strands, guests, stories and contributors. Kate Phillips, 'You'll need to come up with ideas in production. I know that on *Ant and Dec's Saturday Night Takeaway* they spend about three months in pre-production just coming up with all the strands and ideas before they even start filming, they've just got so much that they've got to come up with.'

In this world of reality television, multi-channels and formats – from interactive TV to content for the web, mobile phone as well as TV – ideas are literally consumed so it's useful to make thinking up ideas part of your TV career from the start.

Where do you get ideas from?

TV is ephemeral, it reflects, predicts, picks up and sets the agenda on issues and trends. It can be as ground breaking as creating a format and an entire genre of programming like *Wife Swap* (the first programme to turn a documentary into an entertainment format) or *Big Brother*. You can have ideas for one-off documentaries, observational long-running documentaries or more formatted series which have entertainment and a twist at their heart.

Ideas come from conversations, from newspaper and magazine articles, the radio, YouTube and the internet; from personalities and characters – whether vehicles for presenters, celebrities, contributors or experts. 'It's about how you translate that into TV programmes,' says Kate Phillips. 'I used to see a lot of people who would have read an article about a man who eats a five foot sandwich and they'd say, "Let's make a programme about a man who eats a giant five foot sandwich!" No, no, no! You look at the man who eats a giant five foot sandwich and you think, is that to do with obesity? Do we eat more now? It's translating the article or conversation and putting it into a TV programme. It's not about jumping from one to the other.'

Newspaper and magazine journalism feed ideas into TV, and often stories are picked up by TV companies which then fight to get access to the contributor or story. Says Kate Phillips, 'For example, Dove started doing commercials for toiletries on billboards using real women and now there are two ideas with BBC3 about trying to find real women to front campaigns. And Endemol is doing *Colleen's Real Women* for ITV 2 which is the same idea. Literally, you're driving home and you see that poster, you see that there is a lot of interest in real women and you turn it into a format.'

It's about picking up on adverts, songs, catchphrases, TV dramas and films. When I worked in development at RDF I researched all the films in the US that were a big hit at the US box office. One film I noticed was *School of Rock* starring Jack Black. It hadn't been released in the UK and I hadn't even seen the film but it was a big hit in the States. I researched the movie, then wrote the treatment and turned it into a format. We took it to Channel 4 which gave us money to see which celebrity rock stars would be available and interested in fronting the series. I put together an eight-page list of names and contacted their agents or representatives. Many refused. Many agreed to take part but weren't quite right. I thought Gene Simmons of glam rock band Kiss would be a characterful rocker. I got on with his agent and started negotiations with her. With persistence and tenacity, and after many late night phone conversations to America, I booked him and the executive producer flew to the US to meet him. I never did – but that's telly!

It's also about picking up, changing, and re-working existing formats to make them different or applying the same principles to different areas. Business shows such as *Dragons' Den* and *The Apprentice* (both originally foreign hit formats adapted for the UK market) have spawned many copies, as has *Wife Swap* which not only generated an entire genre of swap programmes but it also broke the mould of programme making by creating factual entertainment or the formatted documentary genre.

It's important to be able to breathe new life into an existing idea and to give it a new spin. It is essential to be as innovative and original as you can. You have to predict trends rather than follow them.

You need to sell your ideas at every stage – to your producer, series producer, exec, production company, contributor and commissioning editor, so it's essential that you are a good communicator both verbally and in writing. You need to have the skills of being a good salesman – charm, confidence and the ability to read other people so you can sense when to stop and when to carry on, to know when to push a point or leave it. You'll also need utter self belief and a refusal to ever give up!

You must be able to speak to people with confidence that inspires trust and ensures clear lines of communication. They must be able to understand what you are asking of them and you must be able to explain an idea or suggestion clearly and simply and in a way that anyone from any walk of life can understand. To translate ideas from initial thoughts to a coherent TV programme you must be able get others to see your vision and share it with you. 'It is absolutely vital to have good written and verbal communication skills,' says Julia Waring.

Do I need to be able to write to work in TV?

The answer is unequivocally, yes. As well as being able to talk fluently and with clarity, learning how to write is a key skill for working in television because it underpins everything – from compiling questions and writing briefs to scripts for shooting and the edit. 'I think being able to write is a big help and is really important, especially in terms of writing a script,' says Daisy Goodwin.

'Writing is not to be underestimated,' says Richard Hopkins. 'Because whether it is pitching a new idea or writing a script or item you need to do it well. Strangely enough, I often say to people it's worth reading *The Sun* and trying to write in that style because I think they write very simple, punchy prose which could almost be read out loud. But learning to write like that is more difficult than it seems. Knowing the basics like spelling and grammar goes without saying. Writing is so important.'

'A researcher will very often have to write local authorities or to punters or to experts, so they must be able to put a letter together without spelling mistakes,' says Julia Waring.

Do I need to study journalism to work in TV?

Being genuinely interested in other people, wanting to ask questions and being fascinated by stories are all important if you want to work in TV. If you are not interested in finding the story and telling it in the most interesting way possible then TV is not for you as this is the heart of the job.

Skills for working in TV are similar to those of being a journalist, such as knowing which questions to ask, when and how; being a good listener; knowing when to listen and when to interject; knowing when to stay silent rather than push for an answer; employing tact and charm to win trust and to get answers. Some of this is intuitive and is based on your own common sense and judgement. You can learn interview techniques and how to edit but at the core of any story is your relationship with your contributors and having a good personal relationship so you get the best out of them.

'You're either born with an inquisitive mind and a "Nosey Nora" character or you're not,' says Liz Mills. 'You have to have a good sense of storytelling, a good sense of choosing a contributor and what makes a person good on TV. Characters are obviously everything in stories. You need to be able to judge and know how to spot a contributor.'

Being able to talk to different types of people, write a brief, get information and access are all useful tools of the trade and these skills can be innate – you might just have these personal qualities but you can learn them by studying journalism. Journalism skills are incredibly useful. I could not have worked in TV without them.

I did a post graduate course at the London College of Printing (now Communication) in Periodical Journalism, a course they still offer. I learnt how to interview someone, write news stories, features, basic law, short hand, and how to find stories and come up with ideas. It's a grounding which has served me well and enabled me to understand how to write a brief as a researcher but also a script as a producer because the key elements are the same.

'TV is journalism, isn't it?' says Peter Bazalgette, former chairman of Endemol. 'I wouldn't make a distinction between print and video or radio, audio. All researchers are journalists really.'

'Having journalism training equips you with essential skills and looks good on your CV. Being trained as a journalist definitely does help,' says Michelle Matherson. 'I never realised how much. So many people ask me to send CVs of people who have been trained as journalists before entering TV. It's because they know how to construct a story and a narrative, and how to get to the heart of a story. I think it's always there but people feel safer taking on someone with that background, especially if they are dealing with the unknown, it inspires confidence and gives faith. Anything to do with writing is important.'

Grant Mansfield agrees, 'I think having basic skills in journalism is incredibly useful. I still look for it. Quite a lot of the skills you use as a reporter or journalist are transferable because you learn about structure and narrative.'

'I got my first break in TV by answering an advert in *The Guardian* to be a trainee researcher at GMTV. I'd worked as a freelance writer on *The Sunday Express* before that, so I had some media experience. My journalistic background definitely helped me get a job at GMTV as it has quite a strong news element to it,' says TV series producer and media lecturer Lucy Reese. 'I think journalistic skills are very valuable – journalism involves research and writing, which are major parts of TV.'

Alex Marengo says he would always hire a former journalist. He says, 'It's great to have someone from a journalist background who understands a story. It counts for an awful lot if someone is able to write a treatment, to supply cogent information in notes from a phone interview. People who don't have that training can be waffly and confusing.'

In my experience the researchers I've worked with who've studied or have some experience in journalism are wiser and a little more clued up than those that haven't and it's given them a head start.

'I think if you study journalism that's fantastic because it's helps,' says Claire Richards. 'A lot of work in TV is about writing, finding stories; about researching and thinking in the same way as a journalist. All those skills are extremely useful.'

But even if you are a trained journalist you need to learn new skills to work in TV as it is a visual medium. 'The major difference between the two is that journalists do not have to get their interviewees to agree to be filmed, turn up to a studio and speak in coherent sound bites. A good journalist can grab a quick interview over the phone and polish the quotes, whereas a TV producer has to deliver a coherent contributor,' says Lucy Reese.

Studying journalism can help you get in and get on and to do your job well. You don't have to study full time, there are short courses, part time or evening courses in writing, interviewing, news or features at respected institutions like the London College of Communication, City University, Middlesex University, Bournemouth University and Leeds University among others. The *Guardian Media Guide* publishes a full list of academic institutions and what courses they offer.

Where should I look for TV training schemes?

The BBC, ITV, Channel 4 and the big indies like Endemol, Shine, Princess Productions and RDF Television all offer graduate/intern schemes. The Producers Alliance, PACT, offers training in different aspects of TV, as do training companies like the Top TV Academy, DV Talent, FT2, and the National Film and Television School.

Skillset is the skills sector council for the creative industries in Britain. It accredits private training companies in a variety of different ways. It has an approvals team which is responsible for accrediting the 17 Media Academies which offer media studies degrees, and which ran a pilot scheme to approve private industry training providers.

'We support training and education providers.' says Skillset's training manager Alice Dudley. 'We are a kind of strategic organisation so even though we do do the odd thing ourselves, we mainly support people who run things.'

Skillset has £1.5 million in its TV Freelance Fund which comes from contributions from by all the UK broadcasters and independent production companies. 'That money is used to invest in training programmes for the benefit of freelancers working in TV,' explains Alice Dudley. 'So we'll put about half the money into New Entrants Schemes and the other half or maybe a bit more into schemes aimed at experienced professionals. The New Entrants Schemes that we invest in tend to be really quite big, quite expensive schemes. So once a year we'll invest in probably ten schemes that will pay six or seven trainees each a year's minimum wage to get on-the-job training. They'll probably get an industry induction for a couple of weeks and then they'll start doing individual placements for about two or three months. They'll probably have two or three placements. During that time they'll be mentored and have some off-the-job training as well to complement what they're learning on the job.'

Director Charles Martin was able to part finance his training in multi-camera directing by applying for funds from Skillset. 'I had already done some multi-camera directing on location but had never directed multi-camera in a studio,' he says. 'I went on a course at the BBC and learnt how to use a camera. Skillset helps fund training courses and I saw that the BBC offered a multi-camera studio director's course, so I applied for it and asked for funding; both were forthcoming.'

But not for the full amount, so Martin asked for help from previous employers. He explains, 'There was a shortfall in the funding of about £1,000, so I wrote to companies who had employed me, asking them to subsidise me on the basis that I would then studio direct for them at a discounted rate. I didn't receive a single response. It was a lesson learnt; however much people flatter you when they want you for a job, you're unfortunately always on your own. I found the extra £1,000 and went on a five day course, it was brilliant. The great thing about courses is you can learn and experiment without the time/cost factor of actually making a TV show.'

'Skillset has done a very good job of subsidising many courses which aren't just about camera operating but also about the responsibility of programme making,' says ITV's Moray Coulter. 'There are many training schemes available and it's free to get on them. There are also in-house training schemes run by lots of broadcasters. The BBC has more training schemes than any other broadcaster. Here at ITV we have a fast track scheme that people both in-house and on short term contracts can apply for.'

Do look at all the main terrestrial broadcasters' websites for up-to-date information about their schemes and courses. I've briefly listed information on each broadcaster at the end of this chapter. Use the media *Guardian* (published on Mondays), *Broadcast* and books like *The Guardian Media Guide* and the PACT directory to find out the names and contact details for all leading production companies. Contact them by email and call to find out what training schemes and work experience programmes they offer. Even if you don't work for them many schemes are open to freelancers.

The Channel 4 website lists different courses and networking events in regions across the UK. The broadcaster runs open days and summer work experience placement schemes.

'We have a work experience policy for 14 to 19-year-olds,' says Jo Taylor. 'We offer people a two week placement at Channel 4. We have an open day for people to come and meet the Channel. We hold a summer school which is a twelve week placement for nine people to work on a particular project. It could be for commissioning, marketing, press or new media. We give them training on our values and also the key skills they will need in the work place.'

Jo oversees forty different projects annually at Channel 4. She says her role is about 'inspiring and supporting creative talent through-out their career, across all genres and making sure everything is multi-platformed; that the programme will work on a number of media platforms like the internet and mobile phones as well as TV. It's about encouraging and creating new opportunities for talent to go into the industry and create opportunities for development when they are in the industry. Channel 4 feels that investing in off-screen talent is core to their public service remit. And I see my role as very much core to that.'

The Top TV Academy runs the Researcher Diversity Scheme for Channel 4 which is funded by Skillset. Every year 18 newcomers from ethnic minority backgrounds get the chance to work as a researcher within a TV production company for 12 months and receive training at Channel 4. The broadcaster also runs other schemes. 'We've got 4Docs, which encourages filmmakers into the industry so they can post their four-minute films.' says Jo Taylor. 'We've also got a skill scheme called Cinema Extreme for writers, producers and directors. We've got two documentary schemes, one called First Cuts with half hours on Channel 4 and More 4, and also a disability scheme called Shooting Party.'

Top TV Academy is a training and recruitment organisation which runs a large number of courses predominately for the independent sector of freelances but it also offers training for broadcasters. 'We also run the New Entrants Scheme for Skillset. We train them for two days a month so that's 10 to 12 training courses a year. We offer those two days per course to the outside world as well; it might be writing treatments and proposals, it might be how to make a taster tape,' says managing director Liz Mills.

'As well as basic courses in researching, it offers a junior researcher course for entry level people who can then join our membership acad-emy. Through the scheme and membership they get network evenings which are hosted by members of the production companies where they can ask about what they do and if they have got jobs up and coming.

'We do master classes including CV master classes, an interview technique master class, whatever we think is appropriate. These master classes are all part of their monthly membership fees.' Top TV Academy also runs a senior researcher course and a director's course, which includes three days editing, taught by people with industry experience.

DV Talent's courses are also taught by active programme-makers. Says Matt Born, 'We think this is particularly important because the industry changes so quickly and it's vital the skills being taught are up-to-date and relevant to the real world of production. We work closely with the tutors in devising the course content, combining their current industry knowledge with our expertise as training providers.'

As well as courses in shooting and editing DV Talent offers courses in production management, sound, pitching and story telling. It also runs many bespoke courses for production companies and broad-casters, but often it is freelancers on limited budgets who pay for their own training. 'Many of the courses are supported by Skillset's TV Freelance Fund which does a great job and provides grants of up to 70% of the cost,' says Matt.

Rather than paying for a training course, try to get on a trainee scheme and learn on the job and get paid in the process. These are often offered by the bigger indies. RDF Television has a long-estab-lished intern scheme. 'We take 16 people a year who we feel are the future of the industry and who have got some real talent about them,' says Julia Waring. 'We give them a good grounding as a junior researcher and then, if we can, we try to keep them in the company if they are really good. The scheme has only has eight people at one time. Their start dates are staggered and they are employed for a period of six months. If everything goes well, they get their first job as either a junior researcher or junior production co-ordinator depending on which route they want to take. That two months is funded by company overheads so that the production is able to sustain a junior person on their team. This gives them a kick start as during those six months those interns get a chance to work on productions.'

Broadcasters' and production company websites are good places to look for intern jobs and work placements. 'What's great about our website is that we post a lot of jobs that perhaps wouldn't be advertised in *The Guardian* or *Broadcast*,' says Jo Taylor of Channel 4. The BBC, ITV and independent production companies advertise their trainee schemes and new job opportunities in the media *Guardian* (published on Mondays and online), *Broadcast Magazine*, the weekly bible of the TV industry, and its website Broadcast Freelancer.

You could also use networking websites like Facebook, MySpace, Headhunt TV and Bebo to contact TV producers and find adverts. Companies like Channel 4 advertise new schemes on these websites. 'We've got a presence on Facebook and a group on Facebook, 4Talent,' says Jo Taylor. 'I belong to Facebook because I think it's really important that people know what I do. I think the traditional way of finding out about jobs is going to change and the way that we market some of our schemes has to change. I'm not one to go down the traditional *Guardian* route. I did an ad recently for a web producer role and I used Facebook.'

'We have traced people we are looking for on Facebook,' says Julia Waring. 'We find a lot of people on it. There are groups on Facebook for different sorts of PDs, researchers and there is an enormous network of telly people who will know each other and they are always each other's friends. So it is useful.'

Having a presence on a networking site means as you progress and become a hot property, people can find you and you can stay in touch with former colleagues, friends and associates.

How do I get into TV?

No matter how old, experienced, well connected or good you think you are you'll need to do work experience. 'You have to expect to start at the bottom,' says Lucy Reese. 'Make lots of tea and smile a lot.'

'When I came out of university we were told we should be aiming at producer level straightaway,' says Claire Richards. It was a ridiculous and uninformed claim. No production company would give a new graduate a job in production without some kind of experience. Degree or no degree, 21 or 31, most people start in TV by doing unpaid work experience then getting a job as a runner. If you want to get a job in TV the best way in is to offer your services for free, though you usually get travel expenses. It gives access into the TV world and an idea of its culture and how it works – you can observe from the sidelines and from this position you can decide whether it is for you.

'When you are really young, you are blind. You follow advice but it just wasn't true,' adds Claire who persevered for a year taking low paid jobs to survive before getting her first break as a production secretary at Leopard Films.

'Nothing beats hands on experience,' says Lucy Reese. 'Get as much work experience as you can to get that first foot in the door. Use any personal contacts, however tenuous.'

'It's all about doing work experience,' says Jo Taylor. 'It's all about using your contacts. If someone comes and talks at your university then follow it up.' Email them and offer your services for free. This will give you a chance to observe the TV environment at first hand, see how it works, gain experience in production and making programmes.

'Contact companies and do work experience,' agrees Michelle Matherson. 'Work experience is where you first get noticed. At the BBC you do more than just make tea. People on placements go out on shoots and work on ideas. It's only a two or three week stint but we want to make sure they get a proper experience and then if they are a runner we recommend them to another company and they could start as a junior researcher. The ones who want to come back and keep plugging away and being persistent are rewarded eventually.'

Using personal contacts is a fast track way into TV. Most jobs are not advertised and you hear about them by word of mouth and from people you know. Researcher David Minchin got his first job in TV doing admin work in the entertainment department at Talkback

Thames on *X Factor*. 'I got that because my sister worked in TV. She sent my CV to people who were looking for someone to do a few weeks' typing numbers into the computer,' says David, who used the introduction to get himself noticed. 'It was very boring stuff but that two weeks work turned into a job as a runner on the *X Factor* as I worked very hard, made lots of tea and kept on smiling!'

Jenny Popplewell also got work experience through her sister who was already working in TV. 'I made one application for work experience and I got it. I hated the first day. I nearly didn't come back. I thought, "I've got a degree, how will I get anyone to notice me or listen to my ideas when I am making tea and running errands?" I was also worried about my mounting debts and the commute to London was crazy. I am so glad that I went back for a second day. By the third day I loved it.'

If you don't know the person but know someone who does, don't be afraid of contacting them. You never know where it might lead. 'I got my first break through a contact,' says Glen Barnard. 'You might call it luck really. I finished my university course and I tried to get work in TV and naively thought it would be easier and quicker than it was. But there were hundreds of people trying to get in. I didn't work for a year and a half but eventually through someone my mum knew I met a woman at the BBC for a chat. I said I wanted a break and did she have any advice? In the space of 15 minutes, she said, "Actually, I like you and your CV. I know someone you can meet, they are looking for runners." ' Through this Glen got his first runner job at the BBC and has since worked at Endemol.

But if you don't have contacts make them! Use any contacts – no matter how tenuous. 'I've given an "in" to my best friend's sisters, or friends of friends and got them work experience. It's a lot easier than knowing no one at all,' says Katie Rawcliffe, head of entertainment series at ITV Productions. Katie got into TV through her sister, Ade Rawcliffe, now a talent manager at Channel 4. Ade hadn't known anyone in TV but got her first break as a researcher by making contacts at Granada when she appeared as a contestant on a game show.

Even if you don't have any contacts at all it is still possible to get into TV with a bit of persistence and lateral thinking. 'I got my first job in TV by going in as a temp,' says Katie Thorogood, head of features and factual at North One Television. 'I really wanted to get into the BBC so I asked them which temping agency they used. I applied to the agency and they placed me at the BBC the next week.' She ended up temping in the acquisitions department before getting a permanent job at the Corporation.

'I applied to every temping agency that specialises in media to get a placement anywhere,' says Puja Verma who got her first work experience placement in TV after trying for six months. 'I sent over 100 emails to companies but had no response at all.' She finally got a work experience placement by subscribing to the StartinTV website.

'Zig Zag Productions found me there and offered me work experience for two weeks for expenses only, but they only paid me half because I was commuting from Farnborough. It cost me £81 a week to commute and they paid £43.' But she took the job for the opportunity it offered.

Her determination, willingness and enthusiasm paid off and got her noticed. And her language skills helped her get her first running job with the company. 'I did the translation on the five documentary, *The Girl With Eight Limbs,* on the rushes tapes, translating from Hindi to English.' Five months after doing work experience she was offered an office runner's job at Zig Zag.

Work experience gives you a chance to make yourself indispensable. 'If you're in a busy office it's a challenge to get noticed when you're doing work experience,' says Jenny Popplewell. 'I approached a series producer who looked really busy and asked if he needed any help. My offer was gladly taken up. If you deliver what is asked of you and more, then people will come back with more requests. He gave me my first break as runner on his production and I have worked with him again and again throughout my career. You need the same enthusiasm for the job that you had when you were first scrambling to get noticed.'

PROFILE

Katie Rawcliffe, head of entertainment series, ITV Productions

Katie Rawcliffe started working in TV while still at university. She became a researcher aged 22, a producer at 24, and an executive producer at 29. She developed and series produced the first series of *Dancing on Ice*, has executive produced the show and acted as a consultant on the format around the world.

She's worked on some of the biggest shows in entertainment including *I'm A Celebrity Get Me Out Of Here* and *Fame Academy*. She became staff at ITV productions in 2008 becoming Head of Entertainment Series. Her role is to oversee the development of entertainment ideas, the *Dancing on Ice* brand and any subsequent specials, spin off series.

Production credits

Executive Producer – *Dancing on Ice 3*. 2007–at the time of writing (spring '09). ITV Productions

Executive Producer – *Hell's Kitchen*. 2007. ITV Productions

Executive Producer – *Holly & Fearne Go Dating*. 2007. ITV Productions

Executive Producer – *Dancing on Ice 2*. 2006–07. ITV Productions

Consultant – *Dancing on Ice*. Australia/Spain/Russia/Netherlands

Series Producer – *Dancing on Ice*. 2005–06. ITV Productions

Series Producer – *Comic Relief Does Fame Academy*. 2005. BBC1

Initial Jan–March 2005. BBC1

Supervising Producer – *I'm A Celebrity . . . Get Me Out Of Here!* Granada Oct–Dec 2004. ITV1

Series Producer – *The Match*. Initial/Endemol. 2004. Sky One

Senior Producer – *Fame Academy*. Endemol. 2003. BBC1
Producer – *Trouble in Paradise*. Endemol UK. 2003. ITV1
Producer – *Animal Park*. Endemol. Jan–Feb 2003. BBC1
Producer – *Ruby*. Princess Productions. Oct–Dec 2002. BBC1
Producer – *Animal Park*. Endemol. Feb–July 2002. BBC1
Producer – *T4*. Channel 4. March 2001–Jan 2002
Assistant Producer – *T4*. Channel 4. Dec 2000–March 2001
Single Camera Directing Course, BBC November 2000
Researcher – *The Generation Game*. BBC. March–Dec 2000
Researcher – *The BBC Hall Of Fame*. Planet 24. Jan–March 2000
Researcher – *California Bay*. Endemol UK. Jul–Nov 1999
Researcher – *Family Pet Rescue*. Endemol UK. March–Jul 1999
Researcher – *This Morning*. Granada Television. Aug 1998–March 1999
Researcher – *Stars In Their Eyes*. Granada Television. May–Aug 1998
Researcher – *You've Been Framed*. Granada Television. Jan–April 1998

What does your job involve on a day to day basis?
I was previously an executive producer within this department and I'm still exec-ing programmes. My job now has increased to trying to get more work into this department, so a lot of it is development. I meet commissioners, talk about what their needs are for their networks, get a sense of what they are looking for at the moment and then work with the development team here, see what we come up with and pitch those ideas.

You've had a meteoric rise to the top – to what do you attribute your success?
This job is staff but I'd always been freelance and I really enjoyed it. I think moving around a lot has helped me. I've got to know how different TV companies work. I've got to work with a variety of different people on a variety of programmes. I didn't get pigeon holed straight away. I've worked on various types of shows.

What was your first job in TV?

I was a clips logger on *You've Been Framed* for a while. That was while I was at university but they employed me for the whole of the holidays and I got paid for it rather than it being just work experience. When I left university I was offered a job as a junior researcher on *Stars In Their Eyes*, this was because I had worked hard, and had made new contacts as a runner.

Did you always know that you wanted to work in TV?

I didn't know what I wanted to do; it was only after doing the work experience.

What made you get work experience?

My sister Ade was a contestant on a game show and then she did some research work on *Stars In Their Eyes*. She was my first initial contact in telly. I was really lucky. She gave me Jane McNaught's name at Granada in Manchester. She gave me the runner work and then offered me my first job as a junior researcher on *Stars In Their Eyes*.

What's been your best TV moment so far?

When *Dancing on Ice*, after much scepticism about it ever being a successful show, received over 11 million viewers on its first transmission, with a 50 per share. It instantly became the highest rating entertainment show that year.

And the worst?

When I worked on *This Morning* as a researcher. I was responsible for finding a performing family. Often you have to find contributors in a very short space of time as most items on *This Morning* are reactive. I phoned several families, and eventually found a family who sounded great and said they really wanted to come on the show and perform on TV, they convinced me they were all talented; they were really up for performing. They came on but they were awful, they didn't speak, they were basically silent. I think they got stage fright. I remember the editor coming up to me afterwards saying 'I think that was possibly the worst item I've ever seen on television!' I was mortified. I think the lesson I learnt was to always try to meet people, do a recce, or ask them to send in some footage, it's much easier these days with more technology such as YouTube etc.

What achievements are you most proud of?

I'm really proud of the success of *Dancing on Ice* because people said it wouldn't work. They said, 'Did you ever see *Ice Warriors?*' That programme didn't work.

What advice would you give to someone wanting to get a job in TV?

Contact and see as many people as possible. You never know when your letter or email will arrive at the right time. Watch TV, it's surprising how many people come for a job interview and haven't even watched the show you are interviewing them for! Also be prepared to do some work experience, and become a runner as quickly as possible. I think being a step ahead is really important. Knowing what people need. Follow what is going on. So, if your job is looking after a presenter make sure you know where they are supposed to be before they know. Make sure you've got, for example, their daily schedule. If you haven't got it, go and ask somebody. Make sure they're there on time. Pre-empt situations where you can.

What skills do you have that have helped you?

My sisters always say I'm really organised. I'm a real planner. I like to know what's happening and when, that way when something unexpected happens you are ready to deal with it. I'm quite straight forward. I don't play games with the people I work with. I'll say from the outset what I want. I'll say I'm sorry when I'm wrong. I'll discuss things with people. I think TV should be collaborative. I always take full responsibility for being in charge of the show but I sit down with my senior team now – the series producers and the casting producers – and we talk through everything together. I'll say, 'What do you think about that?' 'What do you think about this?' Rather than, 'This is what we are going to do'. I don't know everything.

What's your long term plan?
To stay at ITV and help build a successful entertainment portfolio of programmes. And to remain involved in the production process itself, as that's what excites me.

What will you be doing in 10 years time?
I'd like to have found the next *Dancing on Ice*, *X Factor* or *Strictly Come Dancing*.

How do I get work experience?

Draw up an alphabetical hit list of all the TV production companies you want to work with and blitz all them. Pester, haggle and offer to do unpaid work experience. You can find a full list of TV production companies in the *Guardian's Media Guide 2009*, the PACT Handbook or look at the credits of the programmes you like watching and target those companies whose sensibilities you admire.

'Watch the end credits religiously and send fan mail to producers and production companies,' says Lucy Reese. You can also use the *Radio Times* to target executive and series producers as their names are published in the listings. Most TV companies have email addresses on their websites where you can send your CV, but it's also a good idea to find out who the key executives at a company are. Find out who the production managers, series producers and executive producers are. Who are the heads of production? Even though you are offering your services for free, don't expect to get inundated with responses from companies eager to use your unpaid labour. But do persist and keep writing to people – even ones you might have contacted before. The timing may be right the next time, or they might contact you out of the blue and you might get lucky.

Richard Drew got his first job in TV as a runner in 1996. 'I was at Warwick University doing a film and literature degree but I needed to get some practical experience to get into TV,' he says. 'I sent

200 letters to try to get work experience. I remember because it cost me £70 in postage! Out of those 200 letters I had one positive response.

'Mentorn called me and said they had just started a new kids' TV show in Birmingham called *Scratchy and Co*. I went to the interview thinking I was going for work experience and that I was going to be working for free and they said it was for a runner's position. They said, "We're really sorry, we haven't got much money. All we can afford to pay you is £200 a week." I remember being so excited that it was £200 per week! For me it was so much money !'

You can get work experience through training organisations and TV websites. Top TV Academy has close links with TV production companies. 'We do a lot of work experience with companies for our members,' says Liz Mills. 'By the end of a month or two most of them have got jobs as junior researchers or runners.'

You can do work experience even while you are university in your holidays and you don't have to leave it until your last year. If you really want to get a job in TV you should apply as soon as you start university. Katie Rawcliffe did work experience for two years while studying for a degree in Media, Culture and Society at Birmingham University and went straight into a job as a researcher at Granada working on *Stars In Their Eyes* when she graduated.

Susannah Haley, an entertainment researcher who's worked at Granada Television and Monkey Kingdom. She says, 'I got into TV from doing work experience when I was at university,' she says. 'I did English Literature at King's College, so I didn't know how to get in and I didn't know anyone in the industry or anyone who wanted to do the same thing. My university didn't teach media or have a TV station but I was in the drama group, took writing classes and I did work experience in my own time. Hopefully this proved that I was a creative and dynamic person, and it gave me plenty to write on my CV.'

If you are good, work experience can even lead to paid work while you are still a student. 'I did work experience for a company called

West One Television for the three years that I was at university,' says Jo Taylor. 'I wrote a letter and they gave me two weeks' work experience and that lead to me being asked me back for a month. They then said to come back in my Easter break. I worked there every summer for three months. It was paid as I was doing a job.'

What does a production manager do?

Production managers are responsible for managing the money side of a production – overseeing and managing the budget, organising shoots and the edit. They are usually responsible for hiring runners and work experience placements, and in the past were responsible for helping the series producer/executive producer recruit the team – from work experience people and runners to producer directors.

Draw up a hit list of production managers at each company. There might be several depending on the size of the company and each one will oversee more than one show. Production managers are traditionally the first port of call for speculative CVs and specific job applications as they crew up productions (although the emergence of talent finding departments or individuals in independent production companies, tasked solely with finding and recruiting creative production staff is beginning to change this tradition).

Production managers keep files of CVs, sift through them and suggest suitable candidates and arrange interviews. They also negotiate rates of pay in accordance with the budget and they know what is coming up and when, so it's good to establish relationships with them and stay in touch.

Although they are not involved in the creative aspect of the programme, production managers make crucial decisions about hiring and firing, and give references for runners and work experience placements. They feed back their view of how someone has performed which will influence whether that person is taken on and nurtured and earmarked for great things or whether they're let go, or whether they simply stay as a runner and never progress at all.

While runners, researchers, assistant producers, producer directors and series producers are hired on a freelance basis just for the duration of the production, the production manager, especially in smaller independents, is often a permanent member of staff. Even the smallest independent will try to have a permanent head of production or senior production manager who can work with development teams or executive producers costing possible ideas and putting together budgets and schedules for projects once commissioned. It is at this stage that the team works out what staff will be needed on a specific project and for how long.

When you start out in TV production there are two career options to choose between. You can either choose to work in creative TV production making programmes and become a producer or you can concentrate on the money/administrative and logistical side and work in production management.

Production managers are not responsible for editorial content but are there to make sure it is delivered on time and on budget – something which is often a source of conflict between them and the creative staff who want the best programme available without always caring about the bottom line!

Production managers usually start as production secretaries, progressing to becoming production co-ordinators then managers, they may become heads of production and production executives or even run their own companies. Joely Fether joined RDF Television in 1998 as the company's first head of production to oversee and production manage projects such as *Scrapheap Challenge*. The company grew rapidly and she was made chief operating officer in 2003. RDF was floated 2005 and Joely now sits on the PLC board of the group. As the company has expanded she now oversees the eight operating companies within the group.

Production managers are a crucial point of entry who can help you start or stall your career. Always stay in touch with them as they're a useful resource and source of work – they know what is coming up and when, and can give you an idea of when the company will be taking people on.

The Production Managers' Association (PMA) is a forum for production managers and an incredibly useful resource if you want to find out who's done what, and is a professional body of Film, Television, Corporate and Multi-Media Production Managers. All 200 members have a minimum of three years' experience as production manager and at least six broadcast credits (or equivalent). The Association provides a network for both freelance and permanently employed production managers, and information and support for its members, regular social events, workshops and training courses. It produces a list of production managers available and looking for work an its website: www.pma.org.uk

Talent managers/heads of talent

An increasing number of companies, including RDF Media and Endemol, have created the relatively new role of talent manager. This has emerged in recent years specifically to keep CVs and to find and hire creative talent. Talent managers are specifically responsible for hiring, nurturing and retaining production talent like APs, PDs and series producers.

Julia Waring explains, 'Creative resources are talent searchers. RDF was one of the first companies to invest in a database for people working in the industry so that you can interview and track people and see their development. We've got over 7,000 names of people we can call on for appropriate projects on our database. And we know who is in favour at Channel 4.'

There are also talent managers at the following companies:
- Betty
- BBC Children's
- BBC Factual
- BBC Entertainment
- Channel 4

- Endemol (recruiting across it's four companies Cheetah, Brighter Pictures, Initial and Zeppotron)
- Fever Media
- Kudos
- ITV Entertainment
- ITV Factual
- IWC Media
- Optomen
- Princess Productions
- Monkey Kingdom
- Shine
- TalkbackThames
- Tiger Aspect
- Two Four Productions
- Wall to Wall Television
- 20 : 20

You can find out who the talent managers are by looking at the company website or calling them and asking at reception. It's worth dropping them an email and seeing if you can go in and meet them for chat.

'There are so many people out there who you want to keep tabs on; what people are doing and what experience they have had so you can poach them when the time is right!' says Michelle Matherson, talent manager at BBC Factual. 'If someone has worked on something that's fantastic, you want them to come and work again. The people who have the most experience and who people really rate will get jobs wherever they go.'

Most talent managers have a database which keeps up-to-date CVs, photographs, rates, references and availability. It's worth keeping in regular contact with them because although the timing may not be right at one point, things can change and if they remember you they can get back in touch.

'We make notes about people and if we like their programme we get in touch with them,' says Julia Waring. 'We try to interview as many people as possible. Anybody we meet we will put on the database along with their photograph and up-to-date CV which we can keep up-dating as the years go on, and put them up for appropriate programmes.'

Never call in person and ask to see a talent manager at reception. It's too direct and often they will have appointments or meetings. Email them with your CV and ask if they're free to meet – they'll meet you, even if they're not crewing up, and will bear you in mind for the future.

As head of talent at Zig Zag I was a founding member of the Talent Managers' Forum, an informal group of talent finders from different independent production companies. Members recommend good people who are leaving their companies when a production finishes and give references and feedback to each other. It meets regularly to discuss rates, training and other matters.

'It's great to be able to talk about talent issues, to connect with other talent managers and swap talent,' says forum member Michelle Matherson. 'It's nice to say this person has worked with me and they are fantastic but we have nothing else for them and they would be an asset to your next production. That's really useful.'

Here is a list of useful organisations which can help you research, get training and get into TV. I had never been in contact with Skillset or any of the Government funded regional screen agencies until I wrote this book, but they are a good port of call for information, careers advice and information on training.

Useful Organisations
BBC – www.bbc.co.uk
Britain's first and biggest broadcaster, with two terrestrial and five digital channels including CBBC and CBeebies. It makes its own programmes in-house and also commissions independent production companies to make programmes for all its channels. Its main production centre is White City in London but its regional divisions in Cardiff, Belfast, Glasgow, Birmingham, Southampton and Newcastle upon Tyne all make programmes.

It has many work experience schemes including specific ones for people aged 14–17. Most placements are for four weeks and are unpaid; applicants are reminded that they will have to pay for living and travel costs. It also runs several different training schemes. See the website for up-to-date details.

Be on Screen – www.beonscreen.com/uk

Not a recruitment website but a place where companies look for contestants and contributors for their programmes. You can use it find out what's being made by whom and offer your services for work experience or even a story idea. It has listings for entertainment, reality and gameshows and you can get free tickets to see shows and apply to be an extra. All good for networking.

British Academy of Film and Television Arts (BAFTA) – www.bafta.org

The Academy holds four events a month of pre-release screenings of TV and film productions followed by discussions with the cast and crew. It also holds practical workshops, networking events, debates and workshops for children.

British Film Institute – www.bfi.org.uk

The BFI (British Film Institute) promotes understanding and appreciation of Britain's film and television heritage and culture, and runs activities and services including BFI Education which produces a range of resources and training packs for teachers. It hosts conferences, seminars and workshops for people of all ages. The BFI has a film and TV database which has information on over 810,000 film and television titles, 1.2 million people, 210,000 organisations and 23,000 events such as festivals and awards from all over the world dating back to the beginning of film production events.

British Interacitve Media Assocation – www.bima.co.uk

The trade association representing interactive and digital media industries in the UK. It has an searchable on-line directory of members, provides industry news and lists events where there are opportunities to meet leading industry figures.

Broadcasting, Entertainment, Cinematograph & Theatre Union (BECTU) www.bectu.org.uk
BECTU is the broadcast union which represents technical, production and support staff in broadcast, film and independent production. The BECTU learning website is an on-line learning gateway for BECTU members and representatives, designed to help find information about learning of every kind from courses and workshops to active union reps to vocational courses in film, it offers careers information and advice.

Broadcast – www.broadcastnow.co.uk
The weekly journal of the TV industry. Essential reading for knowing who's who and who's making what.

Broadcast Freelancer – www.broadcastfreelancer.com
The jobs website for the TV industry weekly newspaper *Broadcast*, Broadcast Freelancer is a good source of freelance jobs and talent in UK TV and radio. Freelancers subscribe and create their own on-line profile and post their CVs on the site; employers use it to find suitable people. The search engine is a bit clunky, unspecific and does not always yield good search results. It has fewer than 5,000 people on the database and has 200 profile views a day.

Broadcast Freelancer includes a section called the Careers Fairy which contains advice and training suggestions. The site allows browsers to access the archives of previous weeks. Scanning the back catalogue of questions, I found that it has helped a mother back to work after a year out, has given advice to people who have to deal with difficult contributors and advised a journalist on how she can balance her life with her work, as well as advice on how to get into graphics, progress from being a PA, and working in make up. A useful and essential site which is broad in its TV remit.

Channel 4 – www.channel4.co.uk

Channel 4 does not offer production-related work experience as they commission almost all their programmes from independent production companies so they suggest applying to these individually. However, it does offer limited placements. Look for current vacancies under the work experience tab on the left hand side of the website which you can then automatically apply for.

It also runs a number of schemes for new entrants which it advertises on its website, *The Guardian* and social networking sites like Facebook. It has three talent managers who each have a different sphere of influence. http://4careers.channel4.com

Cultural Diversity Network's Mentoring Scheme (CDN) – www.culturaldiversitynetwork.co.uk

Supported by the BBC, ITV, Sky, Channel 4 and PACT, the CDN's Mentoring Scheme is available to help black and minority ethnic (BME) groups* develop a career within broadcasting through a 12-month mentoring scheme with a senior member of the industry. You need at least three years' experience in any area of programme making or broadcasting to qualify for one of the scheme's 20 places.

Cyfle – www.cyfle.co.uk

Cyfle is the training company for the Welsh Television, Film and Interactive Media Industry and is a Skillset (Sector Skills Council) accredited training partner and a national provider for the industry across Wales. It is offers practical training support for professional practitioners through a slate of short courses and practical funding assistance.

* *The programme will be delivered under Clause 37 of Race Relations Act.*
Applications are invited from BME Individuals from the following groups:
black African, black Caribbean, black other, Indian, Bangladeshi, Pakistani, South Asian,
Chinese, Japanese, Korean, East Asian or mixed race.

DV Talent – www.dvtalent.co.uk

DV Talent is an agency representing leading series producers, producer/directors and shooting assistant producers. It is also an independent training provider specialising in a range of practical, industry-focused training courses. It launched a training scheme for series producers, funded by Skillset, in 2008. DV Talent is used by all the major indies and broadcasters for both crewing and training. It is also a kit hire house – specialising in handheld cameras.

Elsa Sharp – talent finder/trainer – www.elsasharp.com

My website has many resources from details of the training courses, tutorials and seminars that I run as well as one-to-one support and downloads of fact sheets with tips on job interviews, CV writing, researching and development.

EM Media – www.em-media.org.uk

Invests in skills development and business consultancies in the East Midlands. It won a National Training Award for its graduate entrepreneurship programme where 40 carefully recruited trainees received 65 days' training, with half undertaking further work-based placements. It supports technical and skills development training with a training provider. Tailor-made training and development, one-to-one training and mentoring support and attendance at festivals and other networking events.

Fast Track (formerly TVYP TV Young People) – www.mgeitf.co.uk/fasttrack

Fast Track is looking for people across the TV industry with ideas, passion and drive. It is a unique event held alongside the Media Guardian Edinburgh TV Festival. It's a chance to network with and gain access to senior industry execs over two days of masterclasses. The delegate pass for the TV festival is free, as is accommodation. If you've been working in any area of TV you must complete an application and be nominated by someone in the industry. The deadline for applications is the end of June.

five – www.five.tv

Five was launched in 1997 and is the fifth national terrestrial analogue television channel to launch in the United Kingdom. Originally called Channel 5, the station rebranded itself as five in 2002. It doesn't make any programmes in-house or provide work experience placements but in their recruitment section it posts their latest jobs. See also www.five.tv/aboutfive/ recruitment

FT2 Film & Television Freelance Training – www.ft2.org.uk

A training provider for people who want to become freelance assistants in the production and technical areas of the UK's film and television industry. Established in 1985, FT2 receives funding from the Skillset TV Freelance Fund and the Skillset Film Skills Fund. It specialises in junior technical and production grades of the film and television industry

Headhunt TV – www.headhunttv.com

Headhunt TV is an on-line recruitment network which was set up in 2008. It's aimed at anyone who works in the television industry. There are specific areas for graduates, work experience, runners and an A–Z of job titles. Employers can use the site for free to advertise vacancies and look at CVs posted by subscribers who pay £100 a year to advertise themselves on the site.

Similar in an operational sense to Facebook, it acts as a network where people who know each other are linked and recommend each other. It's based on the premise that TV is all about who you know.

Once you subscribe you get your own personal profile page, which offers space for a photo and your career highlights – five shows that sell you the best. The profile carries a skills section and users can download showreels and the person's full CV.

Each profile carries three 'recommenders' who are referees. This is the site's unique selling point – new entrants can show the people they've met on work experience, university professors or in their first jobs. Once you've got their permission, the people or 'recommenders' you choose to display on your profile page can give references. The founders believe that being able to display your contacts and share your network of people with others will mean that employers are more likely to believe that you're good at what you do. Networks link to network, you find people through people. You can also search by genre, job title, name.

The Indie Training Fund – www.indietrainingfund.com

The Indie Training Fund is a registered charity to which UK TV and interactive media indies can contribute in return for free courses, in-house training, subsidised trainees for company placements and subsidised training for their freelance workers. ITF runs short courses for industry professionals in business skills, production and people management skills. ITF short courses are open to ITF members, PACT members and freelancers with two years' proven work experience. Previously administered by PACT, the ITF became independent and took over PACT's short course programme and in.indie scheme in 2008. There is a separate website with on-line course booking and an on-line calendar of short courses.

The Inspiring Futures Foundation – www.inspiringfutures.org.uk

Provides careers guidance, information and training for young people aged 15–23. Services are delivered by a team of professionals across the UK and internationally, with the support of partners in business, industry and higher education.

Careers Guidance – this service, called Futurewise, provides continuous support for young people from 15–23 through a combination of personal and online services including one-to-one interviews and an interactive journal.

Skills Development – a programme of courses and events, both in and out of school, which aim to deliver work-related skills to students and allow them to experience first hand what it would be like to work in a particular sector or industry. Follow the links through www.careerscope.co.uk

ITV – www.itv.com
Based on the South Bank in London, ITV is a broadcaster and production company with regional offices. ITV offers 'Traineeships'. These include the News Bursary Scheme, ITV News Group Journalism Traineeship, Foundation Placement Scheme, Enabling Talent, ITV Traineeships and Modern Apprenticeships. These are similar to the graduate schemes that companies in other industries run and normally last between 12–18 months.

It also runs work experience for over 18s still in education but if you are over 18 and out of education you can only be involved in a work-shadowing placement. With this, you can't get fully involved but you can watch what people do in their day-to-day jobs. Find on-line application forms at www.itvjobs.com

Mandy.com
This is a film and TV production directory. You can search for jobs in production as well as acting jobs in film, TV and radio. It has a yellow pages of services including producers, sections on production jobs, vendors and classified ads. Set up in 1995, mandy.com is one of the leading global resources for the production community, receiving 6 million hits a month from 320,000 visitors.

The Media Guardian Edinburgh International Television Festival – www.mgeitf.co.uk/home/mgeitf.aspx
This is an industry event which provides an opportunity for delegates to discuss where UK television is heading and to hear from industry heads and new media pundits what issues are coming to the fore. There are numerous talks and seminars which provide an invaluable opportunity for networking.

Media UK Internet Directory – www.mediauk.com
A media directory for the UK which lists websites, addresses, telephone numbers, live links and more for all areas of the

on-line media, including radio and TV stations, newspapers and magazines. It has a jobs board for entry jobs in TV, radio and journalism, both paid and unpaid.

National Film and Television School – www.nftsfilm-tv.ac.uk
The UK's only film and television school based in its own studios. Students make real films and television programmes in industry-standard facilities.

The NFTS offers short courses, diplomas and MA degrees. It is the UK's centre of excellence for post graduate students who work under the guidance of tutors who are leaders in their field. Established in 1971, the NFTS became a Skillset Screen & Media Academy in 2007 for providing the highest standard of vocational education, training and skills development for film and television, an endorsement seconded by the UK film and TV industry.

National Union of Journalists (NUJ) – www.nuj.org.uk
Some journalists and researchers who work in TV as freelancers belong to the NUJ. It is provides its members with the advice and skills they need to pursue a successful career in the media. It runs a range of skills development courses for journalists including feature writing and screenwriting. There's a full course calendar on the website.

Northern Film and Media – www.northernmedia.org
The regional screen agency for the north east of England, offers a range of services including training advice.

North West Vision – www.northwestvision.co.uk
Provides funding, help, and support to filmmakers and TV producers in north west England. The website includes advice and tips, training and an events calendar.

Producers Alliance for Cinema and Television (PACT) – www.pact.co.uk
PACT serves the film and independent TV sectors and offers training and business advice. Its website has invaluable information

about the industry including advice on working regulations, legal services for new producers, health and safety, and FAQs.

Production Base – www.productionbase.co.uk
The best network for TV jobs. It is a marketplace for hiring talent where freelancers can post their CVs and availability on the Production Base system which can be viewed by prospective employers.

It has the most user-friendly system of all the recruitment websites. I have used it a lot when crewing up productions and searching for talent as it has the most CVs and the least clunky search engines.

As well as the jobs board, Production Base offers a 'watercooler' discussion forum where you can network with other media professionals, and industry news and events. There are varying subscription rates depending on the length and duration of your contract.

Ofcom (formerly the Broadcasting Standards Commission) – www.ofcom.org.uk
This is the statutory body for both standards and fairness in broadcasting. It produces codes of conduct; considers and adjudicates complaints; monitors researches and reports on standards and fairness.

Royal Television Society (RTS) – www.rts.org.uk
The RTS is a charity which provides a forum where all branches of the industry can meet and discuss major issues. It organises conferences, lectures, workshops, masterclasses and award ceremonies. It has regional centres which run their own events and it publishes a useful magazine, *Television*.

RTS Futures – www.rtsfutures.org.uk
RTS Futures is *the* place for young people interested in television. It has an interactive website featuring exclusive content, discussion and debate, where you can build your profile, upload videos and chat to

other members of the online RTS Futures community. It runs free events throughout the year which offer good opportunities to network and get the inside track on all aspects of television. It offers screenings of new TV programmes.

Scottish Screen – www.scottishscreen.com
Scottish Screen is the national development agency for the screen industries in Scotland. It develops, encourages and supports new and existing talent and businesses and offers education and training.

Screen East – www.screeneast.co.uk
Screen East is the screen agency for the east of England. It develops, supports and promotes the film and media industries in the area through four departments: locations, production, audiences and education, and enterprise and skills. It works closely with Skillset, Skills for Media and PACT to provide opportunities for continued skills development. It offers careers advice for new entrants and people already in the industry looking to develop their skills.

Screen West Midlands – www.screenwm.co.uk
Screen WM is a regional agency which supports, promotes and developes a screen media sector in the West Midlands. It develops talent from new entrants and professional freelancers through skills development and training. It provides career advice and access to funding for training.

Screen Yorkshire – www.screenyorkshire.co.uk
Screen Yorkshire is the regional screen agency for Yorkshire and Humberside and provides media training information and support for the region. It offers part funded practical training courses and tailor made packages for freelancers.

South West Screen – www.swscreen.co.uk
Provides a range of services including training information and support for the south west region.

Skillset – www.skillset.org

Skillset is the Sector Skills Council for the creative media industry which includes TV, radio, film, video, animation, photo imaging, publishing, games and interactive media. Partly funded by the creative media industry and the Government, it ensures that the industries remain competitive by making sure that the workforce is as skilled and talented as it can be. It supports skills and training for people in these industries and offers information on jobs and training on its website.

It works with industry, training and education providers, and public agencies to make sure training provision meets industry needs; it accredits courses and approves providers on the basis of quality and relevance to industry. The website provides details of subsidised training schemes; courses and training providers; the Skillset/BFI training course database; support for freelancers and training providers; health and safety training information; and schemes targeted at new entrants to the industries.

Skills For Media – www.skillsformedia.com

The UK media industry's careers advice service, offers a range of services including individual careers advice and a national helpline in partnership with learn direct.

StartinTV – www.startintv.com

StartinTV is for newcomers who want to break into TV. There's free information on the website or you can pay to post your CV which can then be viewed by employers looking for entry level candidates. Runners, researchers, production people and presenters are catered for on StartinTV.

Set up by producer/director David Wheeler in 2001, it offers numerous resources, including CV feedback, valuable weekly newsletters and job alerts, tips on interviews, TV jargon, job descriptions, and one to one support.

Teleproduction – www.teleproduction.co.uk/
An independent freelancer listing and jobs website aimed purely at production crew within the UK television industry. You need over three years' professional TV experience to join.

Telestart – www.telestart.co.uk
The sister website to Teleproduction for TV entrants, it's a free and independent TV freelancer listing and job site aimed at fresh talent and recent graduates in both production and technical fields.

Top TV Academy – www.toptvacademy.co.uk
Established and set up in 2004 by Liz Mills, the former head of factual at Endemol, Top TV Academy is a training organisation which provides training courses for the independent sector for freelancers and broadcasters and works with all the major production companies.

It also helps subscribing freelancers find jobs in the TV industry, working across all production grades and with presenters. If offers different annual subscription rates depending on grade and it prides itself on working with existing TV professionals at all levels and helping juniors/graduates gain entry into the industry. It's exclusive membership scheme provides job opportunities, further training masterclasses and regular networking events where subscribers can meet with top industry professionals. All Top TV Members have their details and CV on the website which is regularly accessed by TV recruiters.

It is a Skillset Academy partner and its courses are accredited by Skillset and approved by an advisory panel of professionals working in the industry.

TV Freelancer – www.tvfreelancers.org.uk
An internet community catering exclusively and specifically to the needs of freelancers in the TV industry. It is unique because it is the only site run by and for TV freelancers and it is independent.

It does not advertise jobs; it can say what it likes – good and bad – about production companies and broadcasters. It campaigns for holiday pay rights, freelancers' employment rights and fair employment conditions.

TV Freelancer started in 2002 as a simple forum on the internet for half a dozen concerned professionals to share views and discuss the problems facing freelancers in the UK TV industry. It is now an active community for those who love making telly. It tapped into a need to reduce the isolation felt by many workers when it came to negotiating fairer contracts and rates, and also began to expose some of the illegal working conditions suffered by a small number of the most vulnerable freelancers. Interactive databases on the website allows users to get up-to-date information about current rates and can also share their experiences of good and bad employers. The mailing list keeps freelancers up-to-date with relevant industry news, and changes which will affect their working lives.

Urban Fox – www.urbanfox.tv
This London-based DV training company offers HDV and DV camera training in one-to-one sessions for freelancers over one or two days, and in-company training on a range of cameras. Run by Christina Fox, an ex-BBC news and current affairs camerawoman, and her husband David, a journalist and PD.

Women in Film and Television – www.wftv.org.uk
It offers national and international networking events, has an annual awards ceremony and offers workshops and training including mentoring and personal development, as well as masterclasses, panel discussions and preview screenings.

PROFILE
Grant Mansfield, chairman of RDF Media Group Content

2007–at the time of writing (spring '09) Chairman, RDF Media Group Content
Chairs the Creative Board for the RDF Media Group. The board is made up of the creative directors of all RDF's UK companies.

2002–2008 Managing director, RDF Television
Oversees as MD of RDF TV including *The Secret Millionaire, Wife Swap, Shipwrecked, Dickinson's Real Deal, Oz and James's Big Wine Adventure, Scrapheap Challenge, Ladette to Lady, Personal Services Required, Don't Forget The Lyrics.*

1999–2002 Director of programmes, Granada Television
Creative and business head of ITV's biggest programme supplier. Was responsible for 2,000 staff, a budget of £146 million per annum and several hundred shows including *Coronation Street, Cold Feet, Stars in Their Eyes, I'm a Celebrity . . . Get Me Out of Here!* and *Tonight with Trevor McDonald.*

1997–1999 Controller of documentaries, features and arts, ITV Network
Was responsible for commissioning ITV's entire factual output (excluding news programmes). During Mansfield's time at the ITV Network, the Channel doubled the amount of factual output in primetime and reversed five years of audience decline. The programmes Mansfield commissioned included *The Second World*

War in Colour, Real Life, The South Bank Show, Airline and . . . *From Hell*, ITV's most popular factual brand.

1995–1997 Managing editor, BBC Documentaries and Features Department
Creative head of department responsible for a wide range of factual output including the *Antiques Roadshow, DIY SOS, Vets in Practice* and *War Walks*.

1991–1997 Executive producer, BBC Documentaries and Features Department
Mansfield devised and produced a number of award-winning and popular television programmes including *Driving School, Holiday Reps, Vet's School* and *War Walks*.

1979–1991 Reporter, Correspondent, Producer, Editor, BBC Television

How did you get your first break?
I started off working in radio and I was offered a job in a news room. It's different now, but then there was a sense that you went down the journalism route be it in newspapers, radio or telly. My background was journalism really, from an honours degree to reporter to on-screen reporter. In television I became a producer.

Did you have a game plan?
No, I didn't really. I knew fairly early on that I didn't want to stay as a reporter because reporters don't have any control over what they are doing which evidently doesn't work for me. So I thought I'd quite like to be a producer. I didn't have a plan when I started off. The main thing was to try to make a living.

What was your first break in TV?
Peter Salmon who was the Head of Digital at the BBC gave me my first job as a producer. He wanted me for a new environmental

programme called *Make Change* because I was considered to be an OK reporter. I said, 'I cannot work on this programme if I can't produce my own items,' and then very quickly I stopped being a reporter and started being a producer. I eventually ended up being the editor.

What qualities have helped you succeed?
Hard work, a certain degree of self confidence and a lot of energy. You need to love the media that you are watching. I've always liked watching telly.

What advice would you give someone starting out?
If you want to work in television you need to watch a lot of it. I think it's quite surprising how many people are fearful when you say to them, 'So, what do you like on television?' That question quite often throws them.

What have been the best moments in your career so far?
There have been lots of good moments. The best move I made was leaving the BBC. I think I had a slightly one dimensional view of the television industry working at the BBC. In the nicest possible way, you become institutionalised. Actually going out into the commercial world where the focus is entirely on making programmes that people want to watch in large numbers was a real eye opener.

What was the worst moment?
Don't ask me that; I think it's pretty obvious! [He refers to the Queengate scandal.] The fall out has been devastating. It's been very difficult for the company and has knocked everybody's confidence. I hope people will realise that we paid a very heavy price. I hope that people will still value the things about the RDF Media Group which they valued before, which is that we're an incredibly creative company and we have some of the smartest, cleverest people working in the industry. They were here before

the handover, they are here now and obviously none of them had anything to do with the editing that lead to our suspension of making programmes. I think we got caught up in a much bigger debate but we made a mistake and we have to hold our hands up to that.

What's the best way to get ahead in television?
Making the right choices is important. I think what drives people mad in this industry and what attracts people into this industry is that people don't stick around in jobs for very long. If you want a career in television you should embrace change. I know a lot of people are frightened by it but I think changing jobs regularly is a good thing. Don't worry too much when broadcasters are restructured because normally when they change, in my experience, it's always been good. It tends to be frightening for a while but a young person starting in television should recognise that they can move around quite a lot. If you don't like change and you want to stay put with one employer then become a lawyer!

Chapter Three
Getting a job

Do I have to do work experience to get in?

The short answer is, yes you do. Work experience placements under-pin the industry, providing an endless stream of free labour from people keen to get into TV. Some companies continuously run placements where people willingly work on productions alongside paid runners and researchers, often doing the same job, but for travel expenses only.

PACT publishes work experience guidelines which offer detailed guidance for companies committed to providing quality work experi-ence placements. In 2006 the Department of Trade and Industry (DTI) and HM Revenue and Customs published guidelines for work experience in the television industry. These guidelines have been co-ordinated by Skillset in consultation with the main employers, trade associations and trade unions. They are intended to offer practical advice rather than definitive statements of law.

'We were asked to develop work experience guidelines for employers by the Government,' says Alice Dudley of Skillset. 'What these guide-lines recommend to employers is that if they have actually given you a job to do, then it's a requirement to be paid the minimum wage whereas if you're doing sort of work experience or shadowing i.e. if you're not expected to turn up at a certain time or leave at a certain time and you're not being given any particular job but seeing what's happening, then it's ok not to pay the minimum wage, if you're in education, for example.'

You can download the guidelines from Skillset's website: www.skillset.org and at PACT on www.pact.co.uk. It's worth reading both sets so you know your rights before you start. However, most work experience people end up working on a production and not getting paid for it. It can be exploitative but if you do the job well it is an opportunity to then get paid work on a production. As much as a company uses you, you can look at work experience as an opportunity to find out as much about TV as you can.

It is an invaluable chance to learn as much as you can about TV production and how it works, to meet new people and network and promote yourself. You might have to do many placements to get work – things don't happen overnight – and the work can be tiring and mundane at times. Although you might feel ready to do research and work on a programme, you will be expected to start at the bottom. Expect to make tea, do boring tasks like delivering tapes, photocopying and getting lunch. But you can use these tasks to your advantage – you'll get to find out who's who at the company and can observe what's going on without any pressure.

If you are really keen to get on, work experience can prove to be an important introduction to the industry and to contacts who will help you secure work in the future. No matter how you feel about being asked to do something always be courteous, helpful and charming. Smile and be eager to please even when asked to do the most mundane or boring of tasks. Never refuse to do something (unless it is completely against your moral conscience) or expect to leave on the dot of six o'clock. You might be asked to deliver a tape at 5.50pm; saying no will not make a good impression. No one likes a clock watcher. In TV you will be expected to work weekends and late nights when required and it's sometimes expected when you're doing work experience. Show willing and eagerness and it will be remembered.

You need to make a good impression during your time at a company. The people you work with – from the receptionist and runners to the production manager and even the MD – will be observing you to see if you have the right attitude and potential. If you do well and impress the people you work with, they'll remember you when the next opportunity for a runner arises and they could offer you the job. People always prefer to hire someone who they know and if you prove yourself during work experience you might be asked back again.

Think of work experience as a kind of probation that works both ways. You're observing each other. While you're at a company find out as much as you can. Take home their programmes to watch, find out who's who in development and find out what is close to being commissioned and might be crewing up soon. Talk to the runners and receptionist about the company and what they think of it – whether it is a good place to work that offers opportunities for progression.

Network, network, network

Once you start your work experience try to speak to and get to know as many people as you can. Make friends with the people you work with, chat with other the runners, the receptionist and other people doing work experience. Be genuinely friendly and offer your help and enthusiasm where it's needed. As you get to know the people you work closely with, ask questions and find out how they've got on.

Don't ask too many questions on the first day or in one hit – you don't want to become a pest. Show a genuine interest in people and talk to them about themselves and their experience. Ask questions when it feels natural and appropriate, like in the pub or at lunch when they're not too busy. Strike up conversations during downtime and get to know people and how the company ticks. Ask other people how they find and hear about work.

Always keep your radar open and switched on; absorb information like a sponge. Notice who is under pressure and offer your help if you have time to give it. Listen to other people's conversation and if they are not too busy ask them what they are doing and if you can assist. Find out who knows who and who's going out with who to avoid any embarrassing faux pas! Chances are they'll talk to each other about you too, so try to make a good impression.

If you are asked to do something and you're not sure how to go about it always ask, though not necessarily the person who asked you! It doesn't mean you're ignorant but that you need clarity and guidance. 'Never be afraid to ask question,' advises Katie Rawclifffe, who has had a meteoric rise in TV becoming an executive producer aged 29. 'No one would expect you to know.'

Repeat the question to make sure you've understood what is being asked of you. Never agree to do something and then don't do it. People are relying on you and if you don't deliver you'll let them down. It won't look good and you'll be unlikely to be asked back. If you feel too embarrassed to ask them to explain, check in with one of the runners, the receptionist, production secretary or one of the researchers who will probably be able to help if they are not too busy.

'You don't have to ask the person who's giving you the instruction,' agrees Katie. 'Ask another researcher. When I started working at *T4*, I was responsible for shooting the start of parts. I had no idea what that meant! I went and asked a cameraman. I was quite honest with him and said I didn't have a clue what to do. He helped me out no end.'

Claire Richards, a shooting AP still asks for help when she needs it. 'You can ask for help and assistance. I have no problems going around talking to production managers or producers and directors and asking their advice, there are so many people I could ask when I need to.'

Once you've finished your placement write personal and sincere thank you notes to the people you've worked with. Thank them for giving you a chance, or for their help and advice. Offer your services again in the future and say you'd love to come back. If you did make a positive impact and they like you, ask them to give you feedback and a written reference. Also ask them if you can use them as a referee on your CV and establish what they will say.

Check that:
- They are happy to give you a reference and for you to give their contact details on your CV or to a third party on request
- It will be a good one
- You know what they are going to say
- What you say you did at the company and what they say you did are the same!

Make sure you maintain your friendships with your contacts and referees – no matter how long ago you worked with them. TV is a long game and you can return the same companies and work with the same people again and again.

TV researcher Susannah Haley did a successful work experience on *This Morning* while still university but made the mistake of not keeping in contact after she left. She says, 'I enjoyed my placement there and then went back to uni for my final year. I didn't realise how many work experience people go through the mill so I didn't harass them to continue working there through holidays like I should have done. By the time I had finished uni, they had forgotten who I was.'

When she left university it was almost like starting again and she had to explore different avenues to get back in. She explains, 'I waited until I graduated and then started looking for work. I didn't have as much luck getting more work experience and I got impatient so I joined Top TV Academy. I did their one day junior researcher course and they got me my first job on *Celebrity Big Brother 3* as a logger. They also organised "networking events" which weren't for me but some people might find them useful.'

Don't expect a placement to immediately lead to a job. Accept that you might have to do more than just one placement before you get a paid work. AP Simon Warrington says, 'I worked at various companies to gain work experience before being offered a runner's position at Eurosport.'

Check in regularly with the company where you did your work experience to hear what's going on and to get ideas of who else might be recruiting. Find out who's commissioning what and where it's being made and target your letters and calls.

How to apply for a job
The hit list
Draw up an alphabetical TV hit list of broadcasters and production companies and update and expand it regularly. Make as many

contacts as you can by targeting production companies and finding out who does what. Start by putting together a list of the companies which are making programmes you like to watch and would love to work on. But your list should be expansive and growing all the time: as your career in TV progresses you will come back to this list and it will grow and develop year on year. As a freelancer you'll refer to it at the end of every contract.

'Your personal network is absolutely vital,' says Moray Coulter. 'Get to know and stay in contact with as many people as possible, you should never presume that people will know that you are looking for work or what you have just been doing. Let people know when you're not working.'

Long-running daytime shows are a good place to get that all important foot in the door. Often they are made by regional companies and are made in volume, because they are on air all the time. They'll have big teams, have several different roles and have a regular turnover of staff which gives a greater scope for getting in and getting promoted. They may not be an immediately obvious choice but daily, topical shows are good places to get a good grounding in TV as they mix TV production with journalism. Contact the production manager/series producer and offer your help as work experience. Find out when the show starts, it's next run and when they are next crewing up. Make a note to contact them nearer the time.

When you are keeping notes for your hit list be methodical, organised and thorough. Make and keep proper notes, never throw away a number and always make a note of someone's name (and how to spell it), their email address and when you spoke to them. It's important to create your own system and to regularly update it, making clear notes of what they said, what you agreed, when you sent them your CV and what plan of action you intend to take, for example whether to call back in a few weeks or days. You can then follow up your call with an email and contact them again in the future. Don't be too keen or too pushy, be persistent but not too intrusive.

You could make up and use a grid like this:

COMPANY	CONTACT AND DETAILS	GENRE/ PROGRAMME	STATUS
Brighter Pictures (part of the Endemol Group)	work.experience@ endemoluk.com runners@endemol uk.com There are two talent managers at Endemol find out who they are and their email and contact number	**Reality/Entertainment** *Big Brother/ Big Brother's Little Brother (C4), Celebrity Scissor Hands (BBC3), You Can't Fire Me I'm Famous (BBC3),*	Emailed [*name*] CV on 21/10/08 to work experience address. If I don't hear email again on 24/10/08
Celador Productions	Head of Production – Email address: Direct line:	**Entertainment** *Jimmy Carr's Commercial Breakdown (BBC1), All Star Mr and Mrs (ITV1), Who Wants to Be A Millionaire? (ITV)*	Sent CV on 22/10/08. Emailed PM to offer to do work experience. No response yet. Chase in a week on 29/10/08

Keep an alphabetical list so as the list-grows you can keep the system in order. Work from this to-do list and keep it up to date. Tick stuff as you go along. It will help you feel as if you are making progress and give you a path to follow.

'I'm quite practical when I'm out of work,' says Richard Drew, who used to use a similar job grid to find work. 'It's important to be really methodical. I had a table with the name of the company, the address, contact number, the name of the contact who I'd contacted there, when I sent them my CV, did I hear a response? Then you just keep following it up basically.'

Don't be afraid to call or contact the same people again and again as things can change within the space of a few days or a week. It's all about being in the right place at the right time and although you might call at a quiet time, they might get a commission and need to put a team together quickly. The next time you call they might have something and you'll be in luck.

Do keep an accurate log of when you contacted them and how (I cannot stress this too much!). The last thing you want to be is a pest and to call too often. Call and use your charm and you might become known to the company.

People never mind the offer of free work experience, especially if you're polite and friendly. Keep in contact with people regularly and find out what's coming up; research their shows and find out what might be recommissioned, what shows they have in development that might be commissioned and when they might be looking to hire in the future.

'I got a lot of work that way because even if you don't end up working at that company you build up a relationship with them,' says Richard Drew. 'Even though I wasn't the right person or it wasn't the right time they started to remember who I was. It's almost a full time job finding work and you've got to take it seriously.'

Always make sure you have the right contact as people can change positions; check before you email by calling up and checking you have spelt their name correctly and that they are still working at the company.

'I think it's really important that you get a contact name for the company that you're approaching,' continues Drew. 'As a very, very last resort I think it's fine to contact personnel if you just can't find a name. A lot of people just blindly find addresses and send off CVs and they might as well throw them in the bin. They're not going to get anywhere. It's worth taking the time to call and find out the name of the person you should be sending your CV to.'

Never send a group email addressing it to several people at the same time. Sending the same email to several different companies at

the same time is bad practice. You should target your approach carefully and make your emails personal to the company and to the person you are writing to – talk about their shows, put in your feedback
and even make suggestions or give a couple of one line ideas.

Compile a list of the type of programmes the company makes and
make an effort to watch them – if you suddenly do get an interview
you'll be asked about them and it's unforgiveable not to know.
Research the company and be aware of its brand, reputation and
output so you know what type of things they make and how they fit
into the broadcasting landscape.

Many people just send out CVs without actually knowing which
company they want to work for or what they produce. There is just no
point in sending out your CV to a company that doesn't do the genre
or show that you want to work in.

Read and watch as much as you can. Go through the *Radio Times*
every week and make a list of the programmes you're interested in
and want to watch. Watch them and make a note of the people who
work on them. Get their contact details and email them offering to
do work experience or offer your services as a runner or ask if they
have time to meet with you for a coffee.

'There were times that I felt down trying to get a break,' said Puja
Verma who graduated in July 2007 and got her first job as a runner in
May 2008. 'I was temping in between working for free and I found that
all the qualities I have – commitment, enthusiasm and communication
skills – were not appreciated. But I was never going to give up. I was
determined. I had faith. I knew it would happen eventually. I did get
low and then I got a job as a runner. It came at just the right time.'

'I sent letter after letter. I sent hundreds, I did four hours every
night for months,' says AP Claire Richards. 'I would write to different
people from going on to websites – Mandy.com, Production Base or
search the Yellow Pages. I also rang around.'

Accept that you might not always get a reply. Even if you do it might
take months and the company may not take up your offer. Don't take
it personally; remain steadfast and dogged in your determination.

You cannot rely on sending out just a handful of letters to selected companies. It is about volume and contacting as many companies as you can.

'It's got to be tons because there are so many people competing for these jobs. There are at least 60 or 70 companies out there making shows. When you're out of work you should be looking for new places because old companies die and new companies spring up all the time,' says executive producer Richard Drew. 'Read the magazines, read websites, surf around. Find out any new company that's got a new commission because if you can get in with a new company at the start it could be a great opportunity. If someone tells me that they've been out of work for two months and I ask how many CVs they've sent out and they say 70 or 80 then I know they're really looking. Some people send out six or seven CVs and they wonder why they're not getting any work. They're waiting for the phone to ring and it's not going to happen. You wouldn't do that if you were researching. You would be pounding the phones. You've got to pound the phones and send your CV out and really go for it.'

You should circulate your CV widely but don't send the same application and CV to everyone. Adapt it each time, highlighting the information that is relevant to the company you are contacting. It is a full time job in itself but different companies have different brands and styles and make different shows which require different skills. Make sure that every application is unique and that you tailor your approach to the style of the company. Target the companies you really want to work for so your interest and passion are genuine. Research your market and decide who you want to work for and why. Start with the places where you really want to work and then increase the size of your trawl as you go along.

'I always say that it's better not to send 100 of the same letter to production companies; you are better off sending 20 that are specifically targeted,' says Jo Taylor, talent manager at Channel 4. 'If you really like the programme, write to the person who produced it. Everyone likes their egos to be massaged and that puts you above somebody else.'

Use a personal approach. As well as emailing your CV you could also deliver it by hand. Leave it with the receptionist to pass on. It's not a good idea to ask to see the person it is intended for, unless they've agreed to meet you for a chat or an interview beforehand.

On a couple of occasions I have been surprised to receive phone calls from the receptionist telling me that someone wanted to give me their CV in person and were waiting for me in reception. I found that a little intrusive. You wouldn't arrive at a stranger's or even a friend's house unannounced and uninvited, so why expect someone you don't know to agree to see you? It didn't impress me. I didn't have time to spend with a complete stranger whose CV and reputation I was unfamiliar with and I cannot imagine a busy producer agreeing to take time out to see someone in that way. They'd need to see your CV, hear of your reputation or have some personal contact to warrant a meeting.

Dropping off your CV and talking to the receptionist can pay dividends though, especially as they might feed back information about the people who matter. 'One way of making yourself stand out from all the other hundreds of thousands of hopefuls is to go into the reception of a company and be incredibly charming to the receptionist. Give them your CV and covering letter and ask to see the talent manager. They'll probably say no. But they will mention you to them and the fact you made the effort in person,' says Jo Taylor. 'When I worked at Optomen I never realised how much power receptionists have because they would say to me, "Oh, he was really rude" or "He's really nice". It might not work all the time but I think it's a little trick you can use if you haven't got contacts.'

How should I write a covering letter?

It's essential that you write a covering letter or email when you send your CV to someone. It's the first thing someone will read and you need to get it right. Get it wrong and people won't even open your CV.

Make sure you get it right as without a good covering letter and CV you're not going to be asked in for a chat. 'We don't advertise. It explains on our website how to contact us and people have got a specific email address to write to,' says Julia Waring of RDF. 'We get a lot of direct mail. We see people who we think are appropriate. The first thing we assess is how someone has written their CV and covering letter. Then we ring them up; it really is on how they sound on the telephone on that first call. It's the decider on whether you are going to bring them in or not.'

You could write your covering letter and add it as an attachment or use the email itself as the covering letter. Jo Taylor feels it is advisable to send a covering letter with the CV as an attachment. 'When I was working at Optomen and the BBC I received CVs without covering letters. People would send an email and use it as a covering letter but they'd do it over a couple of sentences. They'd think that was enough but I think that shows arrogance and laziness.'

Taylor feels that a carefully written covering letter is a crucial way of making an impact. 'I think it shows that you're serious about working in the industry and that you've really researched and thought about why you want to work in it. It shows a courtesy. I just feel very strongly that it's the right thing to do. If I was applying for a job I would always enclose a covering letter. It just shows professionalism.'

In my experience this is an unusual practice and attaching a separate covering letter is rare. Most people use the email as the covering letter and attach their CV to it. Personally, I find this more acceptable than sending a separate attachment.

In whatever form there are golden rules to writing a covering letter:
- Make sure you address it to the right person and you know their title and role
- Make sure you spell their name correctly
- Don't address them too informally – if you are not a friend and you don't know them you are a stranger to them
- Don't be too formal either – it's telly not banking and your style should be punchy and instantly engaging. Humour can help!
- Check and check again for spelling mistakes, grammar and accuracy
- Know the style and brand of the company you are writing to so you can reference it

- Mention their programmes and have an opinion and suggestions
- Suggest a couple of one line ideas that fit their style
- Lead with the most important information first – that which is pertinent to the company and the job you are going for
- Don't make it overly long – keep it brief and to the point – they have your CV to find out more

If you are going to use the same template letter make sure you change the name in the 'To' box and also in the body of the letter. Make changes that are pertinent to the company you are targetting.

I had people emailing me at Zig Zag saying they loved programmes made by other companies, I'd see other company names in the body of the letter. Find out who you are meant to be writing to and make sure that you refer to that company throughout the letter; do not forget to change it.

Julia Waring of RDF adds another note of caution. 'Also have a sensible email address. Not daft names like hotstuff69@hotmail.' Try to get an email address with your name in it, which gives you a chance to become memorable for the right reasons and not anonymous.

People don't have time to read long covering letters. You only have a few sentences to make an impression so make sure you put the most important things first, in a simple but attention-grabbing way. 'Cut from the bottom,' says Lucy Reese. 'This is a newspaper expression that tells you to put the good stuff first. This applies to everything from a job application letter, to a TV proposal, to a programme.'

'If you cannot write a clear covering letter then, again, it's going in the bin,' says Richard Drew. 'How are you going to write a letter to contributors or express yourself to people? I hate really long covering letters. I think you need one to two paragraphs maximum. You shouldn't need more than half a page because the information should all be in your CV. Many people write really clumsy covering letters that just go on and on and are full of gibberish. They try to talk themselves up but they don't actually know what they are saying. They think it sounds smart but it doesn't really.'

Here's an example of a covering letter with several errors.

To: Tom.Wisly@productioncompany.co.uk

From:hotstufftvwannabe@hotmail.com [*silly email address*]

Subject: job application [*non specific and too vague. Which one and where*]

Hi Tom, [*this is too familiar*]

I am writing to apply for the vacancy [*The job role should be specified as they might have several adverts*] advertised on * [*date should also be given*] and have attached my CV for you to look at. [*The CV should have the applicant's name in the title*]

I am hard working, organised and dedicated. Based in Amersham [*This is irrelevant*] I possess a clean driving license [*Incorrect spelling*] and a car. Not only do I have experience in working in Film and Television [*Examples need to be given here*] but I also have a Law degree and ample experience in customer service and administration. [*How is this relevant to the job?*] Having just graduated in July, there are not may [*Spelling mistake*] people who are lucky enough to gain industry experience whilst supporting full time education. [*Not true!*]

Just like the you, I love to explore different formats of programme [*There's no mention the company name or any of their programmes*]. After all, variety is the spice of life! I have worked in digital media, Casting and Promotions, Television and feature films. [*Only list the experience relevant to the job you're applying for. Does this person want to work in film, casting, digital media or TV?*]

Through academic study I have managed to sample the workings of the advertising world and live television.[*More irrelevant information; which did she prefer?*] I have diverse research interests too, from Bollywood to Pornography [*why mention this?*], from Employment Rights to Human Rights. I love getting my teeth into something new. [*All this should be included in a small section of Interests on the CV*]

I am also multi lingual, fluent in French, Arabic and English. [*This should go much earlier*]

I do believe that I will be more than able to fulfil all that is required for this role as I am already familiar with the workings of TV Production [*This sentence is over written and there's no substance to the claim it's making*] combined with my strong administration and interpersonal skills.

I hope you will consider my application. [*This shows lack of confidence*]

Warm wishes, [*Too familiar and informal*]

The person's name
[*There are no contact mobile number or contact details on this letter*]

A better example of a covering letter would be:

To: Tom.Wisly@fivetwo.co.uk

From: elsasharp@*mail.com** [*An email address which includes the person's name*]

Subject: General Runner position at five two advertised on Production Base on 23/08/08 [*Clearly sets out application*]

Dear Tom Wisly

Please find a copy of my CV in response to your advert for a general runner at Five Two Productions on Production Base on 23rd August. [*This exactly states the job. A company might be trying to fill several positions so be clear about what you're applying for*]

I recently completed a work experience at Zeitgist Productions and am currently employed as a cover runner on *The Boy Who'll Never Die* on an ad hoc basis. During my placement I managed to secure access to an important location for free with some gentle persuasion. [*Shows that the person is in employment, has been kept on and has added value to the company by saving money*]

I have a clean driving licence, I know London well and speak French, Arabic as well as English. I can operate a Z1 camera proficiently and I've shot on location as a third camera. I have a showreel you can view. I have also done contributor and location finding. [*All valuable skills to have as a new entrant – good to highlight them in the covering letter*]

I am hard working, keen and enjoy working as part of a team and I am available to start immediately. [*It's always useful to state your availability*]

I enjoy watching Five Two Productions' innovative films including *Freeze Me I'm Yours* and *Love Me Hypnotise Me*. I have some ideas which I feel would suit your brand of programming of shock docs, one of which is a follow up to *Love Me Hypnotise Me* in a 1 × 60 studio based 'stunt' show. [*Shows a knowledge of the company and their programmes and initiative with the suggestion of an idea*]

I'd love the chance to come in and meet with you for a chat. [*Friendly and positive but not too pushy*]

I look forward to hearing from you.

Yours sincerely,

Elsa Sharp
mobile number [*Useful to include two different contact numbers*]
daytime contact number

Do I need to use a gimmick to make my application stand out?

Some people use gimmicks to make their application stand out from the many hundreds that are sent. It can help if you get it right and create something original and impressive; done poorly it can work against you. Most applications will be read briefly and filed. I try to read and answer everyone who has applied to me, but sometimes that is not always possible.

Time is precious in TV and so most applications have just seconds to impress. They'll be quickly scanned so make your application punchy, well written, different and memorable.

Using a gimmick helped researcher Susannah Haley get her first work experience placement. She says, 'I wrote to *This Morning* for work experience with a quirky CV. It was like a kid's book called "The Work Experience Girl" and the blurb on the back was about me. It was a bit of a gamble. Some people probably thought it was naff, but it made me stand out. I sent it to about 10 TV companies but only heard back from *This Morning* who said it really stood out, so it must have struck a chord with them!'

But sometimes they can be ill advised. 'A course tutor told me to come up with a gimmick about myself,' says Claire Richards. 'I produced a vintage bottle of wine with a normal wine label with my name on it and my talents and on the back I listed my skills and what I'd accomplished. It didn't work. I got one interview with a company in Birmingham. I sent in my CV and took in the bottle of wine, but it probably looked like bribery!' And she didn't get the job.

Sending ideas with your covering letter and CV can make your application stand out. 'We advertised for people – from runners to series producers,' says Sue Dulay, a series editor at Ricochet. 'Everyone was saying in their covering letters how driven they were and how they were happy to work long hours. That's a given. Out of 2,000 applications only two people sent ideas, which was really shocking. I thought it was something that was really remiss. Ideas are the currency of the industry.'

Jamie Grayson only had two weeks' work experience at Done and Dusted when he sent his CV around but Julia Dodd, talent manager at Monkey Kingdom was so impressed with his application that she asked him in for a chat. He says, 'Julia told me that she agreed to meet me because my profile stood out from the large number of hopefuls who contact her looking for a start in television.'

Grayson had sent a covering letter, a CV and three sample treatments of different programme ideas which he had devised and written which showed his potential, commitment and creativity.

Writing your CV

A good application is simple, clear, brief and well written, accompanied by a well laid out CV. Your CV should be no more than two pages long – whether you are an executive producer or a recent graduate; if you are just starting out with little TV experience one page is all that's necessary.

Get it right and you've got a foot in the door; get it wrong and you could be consigned to the bin, no matter how perfect you think you are for the position.

There are two key elements to a good CV: presentation and content.

There are many websites and books advising on CV styles and layout. You can even download templates from the internet but so can everyone else, so it won't look impressive or unique. I have seen CVs which have been based on a template which have different sections for education and qualifications, have personal statements and skills; these double up the amount of information and are not useful in TV in my opinion.

You want your CV to be concise, clear and personal to you.

Presentation is key

An ugly, messy CV will not be read and will reflect badly on you. If you can't present yourself well on a CV, you're not going to have the focus to get head in TV and you won't be considered.

'If I get a CV which is filled with a jumble and is badly laid out and all over the shop I think, "What are you trying to say?" When I receive CVs I only really look at the first page; if they don't interest me in that first page then I probably won't see them,' says Jo Taylor talent manager at Channel 4.

So bear in mind these golden rules:

- **Keep the design clear and simple.** Use black ink on white paper in an easily readable font, such as Arial, and try not to use anything smaller than 11 point text. Any smaller and it's difficult to read – no matter now tempted you are to cram as much info as possible. Don't use different colours, fonts or sizes to make it stand out.
- **Put your contact details at the top of each page and make sure they're clear.** You don't have to have your address on every page but make sure you include your name, your mobile phone number and an email address you use regularly on every page. Number the pages.
- **Be concise.** Don't say in 100 words what you can say in 30. Don't pad out your CV. Remember the saying less is more. Don't feel that you need to elaborate your CV to make yourself look more experienced. It's better to tell it as it is. You don't need to list every single job you've ever done unless it directly relates to TV or you have related media work experience. It simply isn't necessary and it shows a lack of clarity and vision. If you've run a TV or radio station, the drama group or magazine at university then do add this and also if you have any film, PR or journalistic experience. And if you have space you can say if you've worked abroad or used transferable skills like selling, interacting with the public, but if you can't get it on one page leave it out.
- **Don't use CV speak.** Use ordinary language that is clear and direct. Don't use what you feel are trendy or professional phrases. Getting your message across in a simple and direct way is key. 'A lot of people use "CV speak" which drives me crazy,' says Richard Drew. 'It's just gibberish. They talk about their intentions and use the language you get from a CV writing book. They use all these sentences that mean nothing.' Write in a clear simple way in a brief but punchy style, rather than using flowery language. Your CV is an important tool to sell yourself and how you come across in the few seconds of reading is critical. Use your CV to list your skills, your achievements and experience succinctly.

- **Be confident.** Be positive about yourself but don't be too over the top, you don't want to come across as being arrogant. Make sure your CV clearly shows what have done. List your strengths but don't claim the glory for a show or a job for yourself if it isn't quite true. 'I think arrogance is quite irritating,' says Kate Phillips. 'There's a fine line between confidence and arrogance'.
- **Make sure there's plenty of white space in the margins so your CV is easy to read and your interviewer has room to make notes.** A clearly written, well laid out CV is a pleasure to read and is easy to scan quickly. Done well it allows the reader to pull out and retain the most important facts. You shouldn't have to look carefully to find them – they should be obvious and stand out. Like most people, I take a printed copy of someone's CV into their interviews and write my own comments and make notes of what they said, so it's useful to have space margins. Anything that is quite hard to read is not going to be looked at twice.
- **Read your CV through at least twice before you send it off to make sure you spot any mistakes.** Even better, get someone else to have a look at it for you. Two pairs of eyes are always better than one; it's often hard to see mistakes in your own work when you're so close to it. 'I used to get other people to look at my CV,' says ITV's Katie Rawcliffe, 'as I am a terrible speller.' There is no excuse for sending out a CV with errors. 'Never send a CV out with spelling mistakes,' says Richard Hopkins of Fever Media. 'The number of people who spell things wrong in CVs is ridiculous.' For Richard Drew poor spelling and typing mistakes are first on his list of don'ts. 'If I get a CV with more than one spelling mistake it goes in the bin. I'm really tough on that because I think if you can't even get your CV right then you're going to be sloppy at work. One spelling mistake is OK, we've all done it. If there is more than one then you're scruffy in your work. I'm really fierce about that. I literally go through them and circulate mistakes. I think it's an indication as to how precise people are. It's your CV, it should be absolutely perfect.'

- **When you print your CV, make sure you only print on one side of the paper.** Use a new sheet of paper for each new page.
- **Don't write a personal mission statement.** You only have two pages to sell yourself and even though you may aim to be an award winning director that's in the future and is not as important as listing your credits, experience and skills. Julia Waring at RDF Television has a database of thousands of CVs; she advises against writing one. 'For me, a personal statement is meaningless because anybody can write one.' People receive many CVs and they want headlines. Your personal statement should be in your covering letter. I don't want that on a CV.' If you do want to sum up your skills start your CV with as short précis but don't call it a mission statement, some people might find that a bit pretentious.

Content

There are two types of CV: chronological and functional. Which one you use depends on your previous work experience and preference. Personally, I favour the chronological CV which is by far the most common. It should always include the following and in this order:

1. **Name**
2. **Personal and contact details.** It's up to you whether you include your address, but your mobile number, email address and home number are essential. An address is not necessary, especially if you don't live in London but are applying to work there.
3. **Useful information/skills.** Useful information includes whether you can drive, if you have a passport with a visa that allows you to film abroad and if you have had any camera training which could give you a break and the edge over another competitor. Being able to drive is essential, especially at entry level.

    As you progress you can use this section to include a summary of your skills. Julia Waring of RDF says, 'It's not worth writing your skills at the top until you're a little bit experienced. For me, if people are more junior I would rather see education at the top, not to see if they got a first from Cambridge, but I want to know when they left full time education. You can immediately see how many years they've been in the industry before you even start looking down the page and seeing what they have done.'

Also include any relevant transferable skills, for example:

- The computer packages you can use
- The cameras you can use
- The languages you can speak
- Can you use PowerPoint?
- Can you write – and what?

Qualifications – include GCSEs, A Levels, your degree and any other qualifications in relevant subjects like writing.

4. **Work experience: going from the most recent job backwards.** In this order list the dates, your job title, the broadcaster and production company followed by a bullet point list of what you actually did on the production and what style of programme it was. Lead with the most recent show you've done, rather than the reverse.

For example:

Mar–Oct 2008 Runner *The Man Who Came In From the Cold* – **Sharp Productions/C4**
As the only runner on this 6 × 60 factual entertainment series my duties involved

- Assisting the PD and AP on location
- Logging tapes on location and ensuring their safe return to the library
- Logging rushes into the library and overseeing their transfer from DV to beta
- Helping arrange travel and transporting the crew, contributors to location
- Assisting the researchers find contributors and locations
- Assisting in casting sessions and looking after contributors

- Some basic telephone research including contributor finding and vetting, fact finding

Jan 2008 Work experience *The Man Who Came In From the Cold* – **Sharp Productions/C4** (two weeks unpaid)

- Worked as cover runner on location
- Assisted the researchers find contributors and locations
- Fact finding and checking
- General office duties

It's not advisable to fudge dates. People always check and when you're found out it's not impressive. Make sure you list dates clearly and make it clear if you did work experience.

'If the dates are left off then it could imply that it was just a week,' says Julia Waring. 'You need to know, especially if somebody is just starting out. There is no point putting just the month and a show title. It could be three days' working as a logger. You should be specific about what you have done and put in brackets if it was paid or unpaid work. Saying "Researcher on *Tomorrow's World*; three months unpaid" is completely different to a paid job. A lot of people do work experience for nothing.'

You can be economical on your CV by leaving out shows you'd rather forget or you weren't proud of; many people do, that's different to exaggerating what you did.

5. **Interests.** Keep them brief no matter how many things you love.
6. **Referees.** Make sure these are people who've worked with you and you know that they will be positive about you. Always make sure they are happy to be a referee.

However, if you are a recent college leaver and/or you haven't got a lot of TV experience you may prefer to go for a functional CV.

A functional CV would include the following:

1. **Personal and contact details**
2. **Skills and abilities**
3. **Achievements**
4. **Education and qualifications**
5. **Employment**
6. **Interests**
7. **Referees.** Use someone who knows you well in an official capacity: a teacher, a boss from part-time work. Make sure they're happy to be your referee.

The benefit of this CV is that it gives you a chance to shout about your unique skills and abilities in a way that your working record can't.

'If I've got any advice on CVs it's to make it impressionistic rather than explicit,' says Moray Coulter. 'We are still tempted to do CVs as a chronological series of jobs we've done which very often doesn't tell an employer what they want to know. They want to know, apart from what programmes you worked on, what did you do? What were you like? What did you get out of it? What did you bring to it? What kind of person are you? It's important that a CV gives a sense of who you are rather than just what you have done.'

Coulter, who founded and ran Production Base before joining ITV adds, 'Some people get hired simply because they worked on a specific programme or they've been recommended by somebody because they know who that person is and can say "I know they'll fit in". Your CV should reveal what your personality is going to be like.'

No matter how inexperienced you are, even if you have no work experience credits, I personally prefer to look at a chronological CV. If you are just starting out as a recent graduate or you don't have much TV experience and don't have much to say, don't worry about having a one-page CV; that's preferable than having to wade through pages of irrelevant information.

Highlight your transferable skills – show where you used communication and writing skills, when you have dealt with difficult people or used persuasion or gained access in different but related professions.

Whichever type of CV you go for, make sure everything you've included is relevant, necessary, punchy and well-written.

Sir Alan Sugar might have famously claimed that everyone lies on their CV but unfortunately if you do lie and get found out it will really work against you. You are likely to get found out and you could end up losing people's trust and maybe even the job.

'The first thing I do when I look through a CV is to question what they have done,' says Michelle Matherson. 'People sometimes have an amazing CV but they actually haven't done what they say they have. I ask them to tell me what they've done and how interested they are on a particular project.' And she, like all people who are hiring, will call their previous employers to check.

The fact is that people do check out your experience and references and not only with the people you've suggested they talk to but other people on the production – the receptionist, other researchers as well as the producer/director and series producer. If you lie you'll get caught out and won't be given a second chance.

'Don't embellish your CV,' warns Kate Phillips. 'Don't bluff if you don't know something. Don't do one of those CVs where you say "I was the producer on this", "I was the director on that".' They're not going to believe it. Don't try and big yourself up, it is much better to be honest and keen. List what you do and what you have done so far."

You'll be thoroughly checked out before being offered a job. A potential employer will want to confirm that you have the experience and skills that you claim, but also how you well you worked with the other members of the team. Kate Phillips adds, 'Employers don't want people who won't muck in. Telly is all about getting on with other people.'

'You have to get extensive references, at least three,' says Julia Waring. These are essential to prove that as well as being skilled and talented, the person is up to the challenges and demands of programme making. 'Anybody can fall to pieces on anything. It depends on how difficult contributors are, what the filming pressures are, how much help you've got. Every job is a new challenge and you are only as good as your last job. That's the precarious nature of the industry.'

A CV should be an organic work that you adapt every time you send it out. You might want to highlight different skills and strengths according to the company you are writing to. 'Don't send a circular CV, send it to one person,' says Richard Hopkins. 'Customise your CV, and bring out the strengths in it to suit that particular company.'

As you progress in TV your CV will get you a job if you have the experience or skills that are relevant to the project that is being crewed up, if you have good references, and have worked on a hit show and will bring something extra to the company.

The following are seven examples of how I think a CV should be written and presented.

1. **A new entrant CV**
2. **A runner's CV**
3. **A researcher's CV**
4. **An assistant producer's CV**
5. **A producer's CV**
6. **A producer/director's CV**
7. **A series producer's CV**

The first is that of a new entrant, someone who has little paid TV experience. I have written this myself using elements of real CVs – I have yet to find a perfect new entrant CV! The other CVs are those of real people working in the industry.

1. New entrant CV

It is possible to write a good CV before you get your first paid job in the industry. By mentioning work experience, highlighting any relevant skills or work you could get a foot in the door.

This is the CV of a ficticious recent graduate who has spent their holidays getting work experience at different companies. Scanning it quickly you can easily glean useful information because:

- It's well laid out
- It's clearly and simply written
- It's easy to read quickly
- It's focused and concise
- It accurately lists the dates and times worked on a production, with the most recent experience first
- It lists other non-TV jobs which have transferable skills

Elsa Sharp

*Mobile: 0000 11 1111 email: elsa_sharp@**.com*

Work Experience
Feb–March 2008 Work experience – Documentary X – Sharp/C4
In this two week unpaid placement on this 1 × 60 ob doc my duties included:

- Running rushes to the office from shoots
- Assisting the production manager
- Ensuring release forms were signed and returned to the office
- Researching facts

Aug 2007 Work experience – Camera Shed – Camera Assistant
In this unpaid placement I worked for a month looking after the camera equipment. I prepared camera hits and went on shoots.

- I trained to use Sony Z1 and assisted on a number of shoots

Dec 2006 Runner/work experience
I worked for a sports broadcaster for three weeks in an unpaid placement.

- Spent time shadowing an assistant producer in the edit suites
- Logged certain sporting events to be used for highlights
- Assisted in the gallery whilst a studio show was being filmed

**Summer 2006 Live production runner/Work experience –
Big Brother 6 – Endemol/C4**
This two week work experience was through my degree. My
duties involved

- Setting up of the location for the viewing public
- Meeting and greeting the executive guests, showing them
  round the studio
- Liaising between the different departments, running errands
- Acting as audience control during broadcast

**Jan–April 2007 Producer & presenter – Anybody's Fair
Game – Fair Game Radio**
I played an active role in this university radio station.

- Researched the topics and cultural events to be featured on
  the show and found guests
- Found rare music which was requested for the show
- Was responsible for cueing the music and controlling the
  sound levels

Skills
Clean driving licence and my own car. Fluent written and spoken
French and Spanish. I can competently shoot a Z1. Showreel
available.
Computer literate: six week course at Business and Secretarial
College (2006). I can use Microsoft Office: Word, Excel,
PowerPoint, Outlook, Access; Lotus Notes; 55 wpm.

Education
2003–2006 2 : 1 BA (Hons) in Media with Spanish from the
University of Anywhere
1997–2002 Secondary School, Northamptonshire, 3 A levels in
English, Spanish, French
10 GCSEs including English, Maths

At a glance you can see that this person:

- Is a self starter, confident and focused having done work experience at four different companies over two years before they graduated
- Has used their initiative to get work experience at different companies on different types of shows
- Has had good training and gained different skills on the job
- Has worked on location-based shows and in the studio on pre-recorded shows
- Has worked with a broadcaster and has experience of live production
- Is used to working as part of a team and has been helpful and supportive
- Can speak two languages fluently
- Got a job as a camera assistant to learn how to use a camera. They've learned how to shoot by going on shoots with experienced cameramen
- Can drive and has their own car
- Is interested in media – from their activity presenting and producing radio shows at their local stations
- Has made the effort to train in computer skills

2 A runner's CV

Here's an example of a good runner's CV. Puja Verma got her first runner's job at the time of writing after applying for several months. Her CV is only one page as she is new to TV. This is all it needs to be.

PUJA VERMA

MOBILE NUMBER
E-MAIL puja_verma@***.co.uk

WORK EXPERIENCE
Hoo Hah Ltd, Runner
Various dates

- Helping on casting days
- Runner duties on shoot days such as maintaining petty cash, managing last minute purchases, arranging lunch orders and booking transportation
- Worked on Veet shoot (27.05.2008) and the casting for the Judge shoot (Casting: 12.06.2008)

Zig Zag Productions, **Office Runner and Translator**
06.05.2008–to date

- Assisting freelance and permanent staff with various production requirements
- Assisting with runs and office duties

Zig Zag Productions, **Translator (Hindi)** 15.04.2008–02.05.2008

- Translating from Hindi to English for the follow up of the documentary *The Girl With 8 Limbs*
- Assisted post production in subtitling and editing

Zig Zag Productions, **Translator (Hindi)** 18.02.2008–03.03.2008

- Checking edits, translations and subtitles for the German and UK versions of *The Girl With 8 Limbs* (Channel 4)

Zig Zag Productions, **Runner** 29.02.2008–01.03.2008

- Running rushes to the office for pilot TV game show *Relentless* (Channel 4)
- Ensuring release forms are signed and returned to the office

Zig Zag Productions, **Runner – Work Experience Placement**
02.01.2008–18.01.2008

- Assisting with runs and office duties
- Researching different and unique ideas for programmes

Cognitive Applications, **Post Production Assistant**
07.04.2006–01.08.2006

- Video editing and image processing

iDream Productions, **UK Extra Coordinator**
01.09.2003–27.09.2003

- Organised and navigated supporting artists on shooting sets for Indian feature film *The King of Bollywood*
- Organised appropriate transportation for SAs and escorted them to locations

REFERENCES
Office Manager at Zig Zag Productions
Production Manager at Zig Zag Productions

PROFILE
- Fluent languages: Hindi, Punjabi, Urdu, English
- Full clean driving licence with car
- British passport
- I can use Microsoft Word, Excel, PowerPoint, Publisher, Adobe Photoshop, Final Cut Pro, iMovie and Adobe Premiere

EDUCATION
2.ii LLB Law University of Sussex, Oct 2003–June 2007
The 6th Form College Farnborough, Sept 2001–July 2003
A levels in Media Studies (grade A), Law (grade B) and English Literature (grade C) and pass in Foreign Languages at Work, French

It is a good example of a runner's CV because:
- It's clearly laid out and easy to read but has personal touches
- She's put her most recent job first and listed her duties on each show
- She's shown that she has experience on a game show and a documentary; at a post production house and a graphics company
- She's provided accurate dates
- She's specified when her work experience was paid

- She's returned to the same company more than once. This shows that she has impressed them while on work experience, they value her and have recognised her worth by offering her paid employment
- She's included other relevant media experience – her time at an edit house and working on a film – which shows her organisational skills and her level of motivation
- She's included the contact details of those she's working with
- It shows tenacity, confidence and drive

3 A researcher's CV

David Minchin is a researcher who's been working in TV since 2006. In this time he has progressed from working as a runner to researcher working at the same company. His CV is two pages long. It's a good example because I feel he's listed his experience concisely and clearly, giving details of what he's done, what skills he's acquired on each job and what he can bring to a production.

DAVID MINCHIN

Home number
Mobile number
E-mail

EXPERIENCE
Jan '08 Development Researcher – Factual Features, TalkbackThames
- Shooting and editing taster tapes
- Brainstorming and researching new programme ideas
- Helping out on other productions within the department including:

Accidental Heroes (BBC1)
- DV operator – drama reconstruction

The Boss Project (C5 Pilot)
- Contributor researcher, DV operator

July–Dec 2007
Junior Researcher – Wish You Were Here . . . Now & Then, TalkbackThames (ITV1)
- Researching contributors, locations, facts, arranging permits
- Shooting GVs on DSR and Z1
- Assisting self shooting PD – Camera assistant/Sound recording
- Managing the schedule and expenses on location

May 2007 Production Runner – The British Soap Awards, ITV Productions (ITV2)
- Setting up time-lapse cameras
- Locating and convincing celebrities to do interviews

Jan–May 2007
Production Runner – WAGs Boutique, TalkbackThames (ITV2)
- Assisting PDs on location
- Shooting actuality/GVs/interviews
- Z1 operator on multi-camera shoots
- Sound and lighting experience
- Edit experience – digitising, finding footage

Feb–Dec 2006
Production Runner – The X Factor, TalkbackThames (ITV1)
- Assisting PDs and crews on location
- Logging and day-to-day support for the production team
- Looking after the celebrity judges and presenters on studio days
- Experience shooting on Z1 and storyfinding

SKILLS

Competent camera operator, with a passion for photography
Broadcast experience on DSR450, Z1, PD150/170
Experienced sound recordist
Lighting experience

Languages
Italian – Fluent
French – Advanced
Spanish – Intermediate

Computer skills
Final Cut Pro – confident with basic features
Avid experience – digitising and basic cutting
I-movie/Windows Movie Maker – fully competent
MS Office–Word/Excel/PowerPoint/Explorer
Graphics programmes – Photoshop, AutoCAD, 3D Studio

Other
Full driving licence. Great knowledge of the London area.
Experienced overseas driver. I lived in Milan for two years and
toured North America, Australasia and south east Asia during my
gap year. I have been on many overseas shoots.

EDUCATION

**2005 BA (Honours) Italian and Design 2 : 1 University College
London**
2003–2005 Politecnico di Milano, Italy

1993–2000 Dulwich College, London SE21
A-levels   Art and Design B, French B, Italian B, General
            Studies B
GCSEs   8 As, 4 Bs

This is a good CV because:
- It's clearly laid out with everything listed in reverse order
- In each role he's listed what skills he's learnt and used to do the job
- He's clearly passionate about shooting. He's learnt to use a camera and his material is being used in different programmes
- In each job he has learnt a new skill and is building up his personal portofolio
- He's worked in production and development. He's worked on different genres of programme from big studio shows like *X Factor* to documentaries on location in the UK and abroad
- He has worked continuously for one production company which means they recognise his talents, he's been good at his job and they value him enough to keep him
- He has found work by reputation from being recommended
- He went from being a runner to a researcher in 18 months on just two productions. This shows he's good and has got noticed
- He lists three referees from three different shows and included their mobile numbers which means he's confident in what they are going to say and is easy to check out
- He speaks two foreign languages fluently and has lived and travelled abroad extensively
- His CV is positive and confident without being arrogant or making false claims

4 An assistant producer's (AP) CV

There are many different types of assistant producer. The role varies according to what genre and type of programme they are working on. An AP in entertainment would probably have a studio-based role which would involve dealing with celebrities, talent and contributors in a studio environment. A factual AP on a documentary would probably never work in a studio at all but would film on location. Their role would probably be finding stories, developing storylines, contributors and locations and supporting the producer/director. They may even shoot a second camera on shoots.

For this example I have chosen Dominic Crofts' CV. He is an experienced assistant producer who has worked on sports, entertainment and factual shows and also in development and as an edit producer.

Dominic Crofts

ASSISTANT PRODUCER
Mob:
Email:

EMPLOYMENT

Apr–June 2008	**Assistant Producer, Rory & Paddy Vs the UK (RDF/five)** Worked on this 'antidote to the 2008 Olympics', which focused on quirky local sports events. Responsible for selecting events, casting, location shooting and script writing.
Aug 2007–Mar 2008	**Assistant Producer, The Alan Titchmarsh Show (Spun Gold/ITV1)** Worked on the first two series of this new daytime chat show. Responsible for creating show items, briefing guests, edit producing VTs and producing on the studio floor.
June–July 2007	**Development Assistant Producer (Zig Zag)** Worked on the development of a 2-hour football preview show. Wrote scripts and

	booked celebrities, liaised with the programme's sponsors, EA Sports.
Apr–May 2007	**Assistant Producer, David Beckham Documentary (19 Management)** Acted as a consultant to 19 Management, writing detailed biogs and questions for interviews with Beckham, Zinedine Zidane, Ruud Van Nistelrooy and Fabio Capello.
Mar–Apr 2007	**Assistant Producer, Laureus Sports Awards (Done & Dusted)** Responsible for producing the Winners' Room presenter, overseeing VTs and creating and directing 'skits' with the awards host, Cuba Gooding Junior.
Jan–Mar 2007	**Assistant Producer, Comic Relief Does Fame Academy On BBC3/Endemol** Produced EVS packages to a tight deadline, worked in an edit producing VTs and wrote script inserts. Produced presenters' voiceovers and booked celebrities.
Dec–Jan 2007	**Edit Assistant Producer, Diary Room Uncut (Endemol/E4)** Edit produced the E4 spin off show, commissioning and editing stories from in the house, as well as writing scripts for voiceover.
July–Oct 2006	**Edit Assistant Producer, Sports Most Loved/Hated (Zig Zag)** Produced two list shows charting the heroes and villains from the world of sport. Responsible for celebrity booking, conducting interviews, writing scripts and working in the edit producing packages.

May–July 2006	**Edit Assistant Producer, Only Fools On Horses on Three (BBC Three)** Worked on Sport Relief's celebrity showjumping programme. Responsible for producing EVS packages to tight deadlines, writing script inserts for the producer, booking and briefing guests and producing presenters on the studio floor.
Feb–May 2006	**Assistant Producer, Bare Naked Lunacy (Bellyache Production/C4)** Helped to produce a comedy pilot presented by Russell Brand and featuring Pritch and Panch of Dirty Sanchez fame. Responsible for sourcing and briefing Eurotrash-style guests, helping to produce location shoots and working closely with the producer in studio.
Jan–Feb 2006	**Assistant Producer, Channel 4 Discussion Show – Drink and Drug Abuse (Atlt Productions)** Worked on a Channel 4 discussion programme on drink and drug abuse presented by Russell Brand. Responsible for finding and briefing all contributors and setting up VT shoots for the team.

RESEARCHER CREDITS

Strictly Come Dancing – It Takes Two (BBC2)
The Match (Sky One)
Big Brother 6 (Channel 4)
Big Brother Launch Week (Channel 4/E4)

Loose Lips (Living TV)
Sky Sports (Archive)
UK Uncovered (also presented)
Attitude (Yorkshire Television)
World's End Television (Development)

This is a good CV because:

- It's clearly laid out with everything listed in reverse order with the dates
- In each role he's listed the skills he's learnt and used to do the job
- You can see a clear progression in his career and you can see where his specialisms are
- He's worked in production and development and has specialised in reality/factual entertainment and sport shows which overlap genres
- He has returned to some companies twice
- It's positive and confident without being arrogant or making false claims
- Having seen earlier versions of his CV, he has cut out that he is a trained journalist, his education and subbed down his early career as a researcher to get everything onto one page
- My only criticism is that he has sometimes omitted the broadcaster.

5 A producer's CV

A producer does different things depending on the genre of programme they are working on. They manage a small team of researchers and APs, and are responsible for content, contributors, ideas and the script. A producer on a studio-based entertainment show would also work in the gallery and in a studio preparing for a recording. A factual producer would probably never see inside a studio but would record shows entirely on location though the tasks and skills required would be similar.

Deborah Kidd is a producer who started by working in drama but has concentrated her career on making specialist factual programmes in science and the arts.

DEBORAH KIDD
Producer/Development Producer

Mobile: **Email address:**

Award-winning producer of factual programmes for all major
broadcasters, including co-productions with US broadcasters
such as Discovery, WNET. I have developed and produced highly
successful, critically-acclaimed documentary series and one-offs,
current affairs series, and popular factual series. Most of the
programmes I have produced have had multiple locations across
the world, many of them in locations that were difficult either
politically or environmentally – all were tightly budgeted, tightly
scheduled, and all brought home successfully.

Credits and employment history
Producer & Development Producer – Feb –July 2008
Off The Fence – Amsterdam
Working across numerous specialist factual programmes
for international & UK broadcasters. Off The Fence is an
international production company with offices across the world.

Development Producer – Nov 2007–Jan 2008 CTVC
Working on numerous specialist factual proposals for all major
broadcasters.

Producer – 'THE INSIDER' Dec 2006–July 2007 Mentorn for C4
Over-seeing development producer for the 20 × 30' current affairs
docs in the two series.

Development Producer – Dec 2005–Oct 2006 CTVC
Working across a number of specialist factual proposals.

Producer/Development Producer – 'THE ROOT OF ALL EVIL?'
April 2005–Oct 2005 IWC Media for Channel 4
I developed this series to commission including co-writing
the script. Professor Richard Dawkins is both a world-renowned

evolutionary biologist and atheist – this series is his journey through the Abrahamic faiths and their effects on the world.

WINNER – BAFTA Scotland – Best Documentary
NOMINATED – Broadcast Awards – Best Documentary Series
NOMINATED – RTS Education Award

Development Producer – Jan–April 2005 IWC Media/C4
As part of a development initiative between IWC Media and C4, to fast-track development producers and for us to have working relationships with the commissioning heads of the main channels, to discuss their channel positioning and current programming needs. I developed a 2 × 1hr series, which was commissioned by C4.

Location Producer – 'ANCIENT SUPERWEAPONS' 2004
Darlow Smithson for Discovery/five
3 × 60' series, building full-scale siege weapons, filmed on location in Morocco.

Producer – 'CRASHES THAT CHANGED FLYING' 2004 Darlow Smithson Productions for WNET/five
1 × 60' on innovations in the aviation industry, filmed in UK, US and Europe.

Development Producer 2003 Darlow Smithson Productions

Producer – 'GREATEST MILITARY CLASHES' 2003 Darlow Smithson for five/Discovery
6 × 30' series filmed in UK, US, Russia and Europe.

Producer – 'SALVAGE SQUAD' 2002 Wall to Wall/C4
10 × 50' series restoring historic pieces of engineering.

Associate Producer – 'THE CAR AS ART' 2002
Zenith Entertainment
60' doc with Brian Sewell on cars as art forms.

Development Producer 2001/2 Darlow Smithson

Location Producer – 'DRIVEN' 2000/1 Ideal World/Darlow
Smithson for C4
2 × Series of this prime-time hit series plus 1 × 60′ travel special
to Monte Carlo and bust!

OTHER CREDITS 1993–1999

AP – 'NOT ALL HOUSES ARE SQUARE'
Tiger Aspect for Channel 4 – 3 × 30′ series on architecture

AP – 'SIEGES'
Scottish TV for ITV – 60′ one-off documentary

AP – 'ROOM AT THE TOP'
Scottish TV's new 30′ live show

AP – 'SUMMER DISCOVERY'
Scottish TV's new daily 2hr live show

PD – 'THE MAKING OF "BADGER" '
Feelgood Fiction for BBC Choice
The first independent commission for BBC Choice – a 30′ doc
following the making of this new BBC drama

AP – 'HOW 2'
Scottish TV for the ITV Network – 3 series and other specials
from 1995–1998

Development AP – Drama and Children's
Scottish TV – various dates

Production Secretary – 'McCALLUM' Scottish TV
Pilot film for a new series starring John Hannah

Film Researcher – BAFTA AWARDS Scottish TV

Prior to working in television, I spent 6 years as a professional
actor in everything from large national tours to fringe shows and

formed a theatre company that won Best New Theatre Company at the Edinburgh Fringe.

Industry Qualifications
Advanced Production Safety, Appointed Person – First Aid

Industry Awards
Winner: BAFTA Scotland for Best Documentary 2006
Nominated: Broadcast, Best Documentary 2006
Nominated: RTS, Education Award 2006
Nominated: BAFTA, Best Children's Factual 2001

Education
Post-Graduate Diploma in Drama – Drama Studio London
BA (Hons) in English and Related Literature – York University

Referees listed

I chose this CV because:
- It's clearly laid out, easy to read and scan quickly with all headings well signposted
- Referees and their contact details were included
- She has summarised her experience in a tightly written paragraph which is concise but sells her skills and experience – including awards and nominations – well
- She's been working in TV since 1993 but despite her long experience she has concisely listed her credits, skills, industry qualifications and awards on two pages. She's done this by condensing her AP credits and listing the shows briefly
- She's worked at different companies more than once and has gone back after freelancing elsewhere which is always a good sign
- My only criticisms are she doesn't list the shows she has developed and she doesn't list her skills.

6 A producer/director's CV

There are many good examples of producer/director CVs. I have chosen Nick Holt's CV for many reasons. He is a self shooting producer director who specialises in making quirky, observational films.

NICK HOLT
Producer/Director

☎
Address

2007 BAFTA Television Craft Nomination: Break-Through Talent (Guys & Dolls)
2007 Grierson Awards: Bloomberg Best Newcomer Nomination (Guys & Dolls)

SKILLS
DV shooting (DSR 570/Z1)
Prolonged filming experience abroad (USA, Russia, Africa)
Secured own commissions
Fellow of the Royal Photographic Society (FRPS)
I-Visa (USA), Russian Visa
Driving and motorbike licence
Hostile environment training
Covert filming

EMPLOYMENT HISTORY

North One TV	Cutting Edge: Storm Junkies (1×60 Channel 4)
Aug–Sept 2008	*Co-Producer/Director*
	Self shooting PD following the UK's most prolific 'storm junkie' Stuart Robinson as he chases the world's deadliest hurricanes and typhoons.
	Exec Prod: [NAME]

Firefly TV Feb–June 2008	**Cutting Edge: 90 Naps A Day (1×60 Channel 4)** *Producer/Director* Self shooting PD following a group of British narcoleptics attending the annual Narcolepsy Network Conference in the US. <div align="right">Exec Prod: [NAME]</div>
Minnow Films Nov–Dec 2007	**3 Minute Wonder: 12 Films of Christmas (2 of 12×3 Channel 4)** *Producer/Director* Self shooting PD following a sprout farmer and remote postman as they prepare for Christmas. <div align="right">Exec Prod: [NAME]</div>
Special Edition Films July–Oct 2007	**The Mentalists (1×60 five)** *Producer/Director* Self shooting PD following various competitors during the World Memory Championships 2007. <div align="right">Exec Prod: [NAME]</div>
Close-Up Films March 2007	**Cutting Edge: Phone Rage (1×60 Channel 4)** *Co-Director South Africa* Self shooting PD across the South African part of this film exploring our relationship with customer service call centres. Dir: [NAME] Exec Prod: [NAME]
North One Television May 2006–May 2007	**Guys And Dolls (1×60 five)** *Producer/Director* Self shooting PD on this observational documentary exploring the relationship between men and their life size synthetic partners. <div align="right">Exec Prod: [NAME]</div>

Betty TV Jan–May 2006	**Diary of a Mail Order Bride (1×60 C4)** *Associate Producer* Shooting AP on this observational documentary following the personal journey of a mail order bride from Russia into the UK. Dir: [NAME] Exec Prod: [NAME]
Blast! Films Sept 2005–Jan 2006	**Dark Side of Amateur Porn (1×60 C4)** *Associate Producer* Shooting AP on this one off about the perils of becoming involved in the UK amateur porn industry. Dir: [NAME] Exec Prod: [NAME]
Betty TV Nov 2004– Sept 2005	**Supersize Surgery (7×30 ITV)** *Associate Producer* Shooting AP across this year long observational series following patients undergoing surgical treatment at the country's only private weight loss hospital. Exec Prod: [NAME]
LWT Factual April–Nov 2004	**Britain's Toughest Pubs/Seaside Resorts (5×60 Sky One)** *Associate Producer* Shot 2nd unit camera on DSR500 across this prime time Sky series exploring the country's toughest pubs and resorts and the underground characters that run them. Series Prod: [NAME] Exec Prod: [NAME]
Hart Ryan Feb–April 2004	**Behind the Crime: 'Loads of Money' (3×60 C4)** *Associate Producer* Found and negotiated access to two of the country's most successful counterfeiters.

Series told their stories and that of those who caught them.

Dir: [NAME]   Exec Prod: [NAME]

Nautilus Films
Sept–Dec 2003

Crime Scene Academy (6×30 five)
Associate Producer
Shooting AP across this series following students through America's ground breaking 'National Forensic Academy', an intensive practical course that includes the notorious Body Farm.

Exec Prod: [NAME]

Lion Television
Nov 2002–
Sept 2003

The Slot: 'Compatible' (3×3 C4)
Producer/Director
Produced and directed this series of short films exploring the relationship between man and machine. Own commission.

Exec Prod: [NAME]

The Real CSI (4×60 five)
Associate Producer
AP spending 4 months following the Las Vegas Police Department. Involved sensitive filming and difficult access.

Dir: [NAME]   Exec Prod: [NAME]

House Trapped (3×60 C4)
Associate Producer
Assistant producer on this series that looked at the underside of the property market.

Dir: [NAME]   Exec Prod: [NAME]

Uden Associates
Nov 2001–
Sept 2002

Mysteries of the Ancients (8×30 five)
Associate Producer
Series that revealed the truth behind some of the worlds most

trusted history. Presented by Bettany
Hughes.

Exec Prod: [NAME]

Discovery Europe
Editing and scripting various documentaries
including a 5 part series exploring Las Vegas
(*Sex, Dice and Vegas*), an investigation
into cloning (*Understanding Cloning*), and
an in depth look at the Vatican (*The Papal
Order*).

BBC
Aug–Nov 2001

Watchdog Parliamentary Special (1×30 BBC1)
Researcher
One off special looking into MP expense
accounts.

UK's Worst (2×30 BBC1) *Researcher*
Looking into UK's worst 'Restaurants' and
'New Homes'. Involved covert filming.

**Peter Williams
Television**
July–Aug 2001

Development Researcher
Development researcher exposing the system
of abuse among NHS gynaecologists as part of
**Panorama 'Carry on Dr. Neale' 1999 (1×60
BBC1).**

The First Human Clone (3×30 Channel 4)
Development researcher gaining access to
Italian Professor Severino Antinori in his
race to clone the first human being.

Channel 4 Sheffield Documentary Pitch Finalist 2002
Follow Thy Leader – Researched, wrote and presented this
pitch to a Channel 4 panel about ambitious young Tory
politicians.

I chose this CV because:

- He's listed his awards first followed by his skills
- He has summarised his skills before his work experience which is useful as it's important that he can self shoot, has an I-Visa and has filmed abroad
- He's then listed his credits in reverse order – these are an impressive array of prime time documentaries where he has been the sole producer/director
- His CV shows a clear career path in factual and observational documentaries where he has secured access to difficult subjects
- He's moved around working for good quality production companies, working firstly as an AP and then as producer/director on popular, interesting films for terrestrial broadcasters
- It's clearly laid out, easy to read and scan quickly, with all headings well signposted
- He listed a confident six referees and he's listed the executives on each programme he's made which makes it easy to get additional references and shows he is happy for people do so

7 A series producer's CV

Victoria Ashbourne is a successful entertainment series producer. This a good example of a well laid out CV with impressive programme credits.

Victoria Ashbourne
Contact:
E-mail:

Series Producer
My Little Soldier   A pilot for a new Saturday night
                          entertainment show for ITV1.
         May–Aug 08   Executive Producer: [NAME]

Series Producer
All Star Mr and Mrs   Saturday night entertainment
                          show for ITV1.
         Jan–April 08   Executive Producer: [NAME]

Series Producer
DEVELOPMENT   Running the entertainment
                          development team whilst also
         July–Dec 07   developing All Star Mr and Mrs
                          for ITV1.
Head of Entertainment: [NAME]

Series Producer
The Dame Edna Treatment   New Saturday night
                              entertainment series for ITV1.
         Jan–May 07   Executive Producer: [NAME]

Series Producer

The New Paul O'Grady Show Series 1 & 2 following Paul's
move to Channel 4.

Jan 06–Jan 07 Running a 55 strong production
team.
Executive Producer: [NAME]

Series Producer

Generation Fame Christmas special remake of
'The Generation Game'.

Oct–Dec 05 Executive Producer: [NAME]

Series Producer

He's Having a Baby A new series for Saturday nights
on BBC1.

June 05–Sept 05 Executive Producer: [NAME]

Producer

FAQ U A late night comedy discussion
show for Channel 4.

April–May 05 Featuring Justin Lee Collins,
Ian Lee and David Mitchell.
Executive Producer: [NAME]

Producer

The F Word Developed a new food
entertainment show for C4
hosted by Gordon Ramsay.

Feb–April 05 Executive Producer: [NAME]

Development Producer

Jan–Feb 05   In-house development.

Producer

Celebrity BBLB   A reactive daily late night
live show on Channel 4
hosted by
Dec 04–Jan 05   Dermot O'Leary.
Executive Producer: [NAME]

Producer

The Paul O'Grady Show   A live hour-long daily chat show
on ITV.
Oct–Dec 04   Executive Producer: [NAME]

Producer

The Graham Norton Effect   Based in New York producing
'Comedy Central's' first
June–Sept 04   ever hour long show.
Executive Producer: [NAME]

Producer

Graham Norton Pilot   Developed and produced
a Saturday night
entertainment
April–May 04   show for BBC.
Executive Producer: [NAME]

Producer
NY Graham Norton   Based in New York producing an
                             hour long Friday night show
Jan–Mar 04   for Channel 4.
                             Executive Producer: [NAME]

Producer
V Graham Norton   Series 3 & 4. Included producing
                             a week of Christmas shows
Jan–Dec 03   from Los Angeles.
                             Executive Producer: [NAME]

Associate Producer
V Graham Norton   Ideas, editing, and setting up
                             surprise phone calls and
                             webcams.
Sept–Dec 02   Executive Producer: [NAME]

Associate Producer
Britain's Sexiest   Responsible for finding Britain's
                             sexiest contestants.
Aug–Sept 02   Executive Producer: [NAME]

Associate Producer
V Graham Norton   Series 1. Co-ordinating guest
                             research and interviews.
April–July 02   Executive Producer: [NAME]

Senior Researcher	
So Graham Norton	Finding props, websites and general show ideas.
Oct 01–Mar 02	Executive Producer: [NAME]
Senior Researcher	
Lily Live	Responsible for booking and researching celebrity guests.
July–Sept 01	Executive Producer: [NAME]

I chose this CV because:

- It's very clear and easy to read
- She's pursued a clear career path and has a good track record in successful, primetime and live entertainment shows working at some companies several times and with the same executive producer and presenter which is a good sign that she is likeable, popular and can do a good job
- She's listed the Executive Producers on all her shows and has worked with several many times. One has hired her repeatedly since 2001
- She's managed to keep her CV to two pages by listing all the salient points but without a huge amount of detail, she's omitted her education but at this level it's not that important
- The only criticism is that it would be useful to know what she's done on each show but as most are prime time big entertainment shows many of the skills and duties would be the same and the CV could become repetitive

No matter what your grade or experience or which format you send your CV in, the most important thing you should do is to label your CV. Include your full name and save it as a document that includes your grade and the date. For example,

Elsa Sharp – Producer/Director CV – September 2009

or even briefer and include a bit more detail

Elsa Sharp – Self shooting PD CV – 12.09.08

This makes it easy to save, easy to find and says what it is. An incredible number of people – even very experienced ones – send their CV as 'My CV' or 'CV1' or as just their first name or initials. This is unprofessional and makes it easy for it to get lost in all the many other thousands received. Label it and make sure it's clear.

What are the golden rules of interviews?

From the minute you walk through the door in an interview you are being assessed – even before you've said a word. From how you walk, what you're wearing, whether your body language is confident and how at ease you are, the interviewer is looking at signs of non-verbal communication. We all judge others on first impressions and the first 10 seconds is said by psychologists to be the make or break time when impressions are formed.

'First impressions go a long way,' says Jo Taylor of Channel 4. 'I always think you should look people in the eye. A strong hand-shake is always good at the beginning. It is true that, unfortunately, people make opinions about you before you've even opened your mouth.'

Interviews can be nerve wracking, especially if you're not used to them. Try to be calm and confident. When you first walk in the room try to ensure your body language is confident even if you're not feeling it. Look the person you're meeting in the eye, shake their hand and smile. Be polite and friendly, walk tall. Be charming, the most important thing is to make sure they like you – they'll be judging you not only on your skills, experience and knowledge but also on your personality and will be assessing what kind of person you are and how you might fit in with a team and get along with the people already working there.

'I think you need to have a certain personality to work with certain teams and certain programming,' says Moray Coulter. 'Key to finding the right person for the job is whether they are going to fit in. If you are a particularly demanding senior producer you will need people to work with you who are robust enough to be able to work under pressure without taking it personally or walking out on the job or so lazy that they do a bad job. Similarly, if you are a brilliant creative then you may want someone who you can rely on to be organised and get the job done. That may matter to you more than their creative ability. Or you might be looking for someone who has that bright creative spark to bring light to a project.'

It's good to bear this in mind throughout your career in TV because as a freelancer you'll have lots of interviews and you might not get the job – not because you're not talented or skilled but because your personality might not blend in with the needs of the existing team.

When you are meeting someone you've not worked with before it's important to establish a rapport with them. You want to them to like you – if you do that you've got over the first and most important hurdle already.

Claire Richards got her first break by winning over the people who interviewed her. 'I sent so many CVs and rang around; I eventually got an interview the day before Christmas Eve,' she says. 'I came down from Preston by bus, I had an interview with the production manager and the production co-ordinator. I got on with them instantly. I've stayed friends with them now.'

While you can't always guarantee that someone will like you – nor you them – there are certain things that you can do to make sure your interview goes well.

The dos and don'ts of interviews
DO
Arrive on time
Being late gives a bad impression. 'Lateness is a real no, no,' says Richard Drew. 'But I could forgive if someone calls and says "I'm really sorry, I'm running late" then I don't mind so much.'

If you are late, make sure you apologise when you meet the person, rather than to ignore it. You might want to move on and not mention it but it's bad manners.

'Worse is if someone turns up late and doesn't even apologise. They may as well go home because they are not getting the job,' says Richard Drew.

Try to explain why, otherwise not only may your interviewer think you are rude if you're late but they might think that you are a poor time keeper generally and will be late on the production, even if the events were beyond your control.

It's all in the planning. Make sure you know exactly where you are going. Work out how long it will take you and make sure you allow plenty of time for travel hold ups. Work backwards from your appointment time and work out how much time you'll need to give each leg of your journey.

Know who you are seeing and for what job/position
Make sure you know who you are seeing – is it the talent manager, a production manager or a producer on the programme? Is it a general chat or a job on a specific production? Is it a work experience placement or a runner's job? What's the show? Who's it for? Being well briefed before the interview will give you confidence and will help you answer the questions put to you. If you can't find out before you get there – ask the receptionist.

Dress appropriately
TV is not a formal industry, you're not expected to wear shirts, suits and shiny shoes, but there is an unwritten code. It's an industry based

on image and so how you look is important. Most people wear jeans and trainers to work, especially on shoots and in studios, but the gear is often designer and expensive. You don't have to be achingly trendy, but don't be over smart or scruffy.

'Dress well, smart, casual,' suggests Richard Hopkins. 'But don't overdo it, don't come in dressed in a suit and tie, we don't do that in TV. But there's no harm in coming in wearing a jacket and an open shirt and a decent pair of trousers. I know you wouldn't normally wear that but if you can't take the time and effort to put on something half decent then I would probably think you're a bit of a disorganised slob.

'Don't wear sexy clothing, but don't dress like a nun or a vicar,' he adds. 'They always say when you pitch a show, you should wear "neutral" clothing. Don't wear clothing that is going to detract from what you're saying, which I think is quite a good rule of thumb really. You can wear hip clothes, but try to make it subtle. You can make a statement without making it too loudly.'

It's about striking a balance of wearing smart but stylish, clean and pressed clothes. 'Some people have a tendency to be really quite scruffy and I don't think it's too impressive,' says Richard Drew. 'One girl turned up in knee high boots like ones Julia Roberts wore in *Pretty Woman* and a really short skirt. I thought, "Why are you wearing *that* for an interview?" It's terrible because it's inappropriate and it makes you judge them. Another person looked like they had just been running and their shirt was creased. He was really sweaty and he looked a mess. It made him look really disorganised.'

Be prepared

How prepared you are can make a crucial difference to your confidence and your ability to answer questions. Being well prepared means that you'll have an idea of what the company might be looking for and you can prepare yourself to answer that need.

Before she worked at Channel 4, Jo Taylor had a long career in TV production and worked for several different production companies.

She says, 'I'd always really research the people I was going to see. You have the time to do that. It's about giving yourself the time to prepare. People really notice when candidates come in and they have prepared. It shows. I get much less nervous when I know about the company.'

Watch TV

Make sure you go to an interview knowing what the company is, what they make and what you think of their programmes. 'There needs to be a sense that somebody has thought it through,' says Grant Mansfield. 'People need to do their homework. Turning up unprepared, thinking that they're just going to get through it on their wits alone is not going to get them the job.'

Make sure that you're watching what's hot on TV so you have an opinion and can contribute your views in an informed way. 'I'd come along with a clear sense of what channels I like; what I like on television and what I didn't like and why,' adds Mansfield. 'The truth is, there are several hundred other people who are just a clever as you. It's about trying to emphasise a point of difference. This industry is trying to appeal to 16 to 24 year olds and so if you're 22 or 23 years old you may have a view of the world or an idea which will appeal to that very demographic. So you'll get respect because you're talking to someone that is older than that demographic.'

'I would be impressed by someone I could have a conversation with at a deeper level than I would expect from a new entrant,' says Jo Taylor. 'Someone who watched a lot of telly and you could ask their opinions about for example, *The Secret Millionaire*, and they can come back with interesting insights. Or someone who can show in their work experience or college course the area they want to work in. This shows a commitment, enthusiasm and quite a strong idea of where they want to go but they weren't so set in their ways. You want them also to be open and malleable. That's what I hated when people would come in and say, "Oh, I only want to work in this particular genre". You've got to move around; the greatest thing is to have experience.'

Know the company you're going to see

Know their brand, watch their programmes and have an opinion about them. 'It's very important for people to take the time to find out about your company and the job they're going for,' says Richard Hopkins. 'Research the company you're going to and the shows they've done in the past. If you have made the effort to watch shows they have made and you know what you like and don't like about them, try to limit the amount you don't like about them! They will very much take to someone who has watched their shows and likes them!'

Because, of course, everyone loves flattery; used sincerely it will go a long way.

Take ideas and notes

As well as enjoying flattery, people love ideas even more – make sure you take in ideas and suggestions. As I have mentioned before, ideas are a currency in TV, if you have lots of them and you are good at thinking them up you'll do well. Suggest new spins on the company's existing shows, come up with a couple of items or subjects for the programme to explore.

If you are serious about working for a company you have to take ideas. They are a pre-requisite. 'Come in with a couple of ideas,' says Richard Hopkins. 'Even if they're not right, at least your putting your best foot forward and giving it a go.'

Anna Blue got her first role as a researcher on *The Big Breakfast* by going to the interview armed with ideas. 'I was working as a junior researcher at BBC Birmingham at the time but that had been my first job so I had very little experience. I desperately wanted to work on *The Big Breakfast* and so I wrote a fairly standard letter to HR and got myself an interview. I had a note book with me in which I had sketched various ideas for end of show games; they were cartoon like drawings of contraptions and sets,' she explains. 'When I first started going to job interviews I treated them like exams, I really thought you weren't allowed to take anything in with you. Then someone tipped me off that it was always good to go to an interview with a book of ideas and a copy of that day's newspaper under your arm but make sure you've actually read it. So I spent ages filling up a hard back book with ideas and then casually flicking through it in the interview. Anyway, it must have worked because I got a researcher's job and I ended up staying for three years, although I don't think any of the ideas in that book ever made it to air!'

Prepare for the questions they might ask you and know what you might answer

Although you can be creative in terms of the ideas that you take into an interview, do swot up on the questions they might ask as if you were studying. 'It is like doing an exam,' says Richard Hopkins. 'Just do your research, know your facts and write down a list of questions you'd least like to get asked and think about what the most cunning answers would be to those questions.'

Draw up a list of things you think they might ask you and what you might answer. The more you prepare the more confident and relaxed you'll feel. Know what relevant points you want to get across so that you can incorporate them into your answers and put yourself in the best light.

Expect to be asked things like:
- What are you up to at the moment?
- What does your job involve?
- How long have you been doing it?
- Why do you want to work in TV?
- Why is it that you enjoy entertainment TV?
- Which TV shows do you watch?
- Which TV shows are your favourites?
- Which TV show might you like to know what goes on behind the scenes?
- When could you start?
- Could you tell us two TV ideas that you came up with for a BBC3 entertainment show and an ITV Saturday night show?

- What do you know about us?
- Why do you want to work here?
- What do you think you could bring to the company?
- Why should we hire you?
- Who are your referees?

Have spare printed copies of your CV to give your interviewer(s)
It's useful for them and shows that you're organised.

Take a show reel if you can shoot
Even at entry level or if you've just left university, it could make the difference to you getting the job or not. An extra camera and a skilled person to shoot adds value to a production.

Take in thank you letters and references
As a researcher I had thank you letters from every single series producer I worked with. They're a good way of gauging what someone's going to say about you and is another string to your bow in an interview. Take in references if you have them to leave with your interviewer.

Have the contact details of your referees to hand
Be ready to hand your referees' details over – if you're asked for them in the interview it's a good sign, so be ready!

Appear really keen
Even if you think about it afterwards and are not so keen or you prefer something else you've applied for – appear really keen. You want to be in a position to be offered the job; you can consider your options later, once it's been offered to you. Don't dismiss anything – you may not get offered another position and you should take opportunities as they are presented to you.

Ask questions

Have a list of questions that you want to ask about the show, the team or the company. Find out as much as you can and use the interview as an opportunity to find out more about the company, it's style, culture and personnel. 'When I go for a job I actually think, "Do I really want to work for this company?" ' says Jo Taylor. 'When starting out, people tend to be quite desperate to get in because they really want to do it. You don't always think, "Is it the best company for me?" It's very hard and it takes courage. It's a good tip for when you go for an interview. It makes me less nervous to think, "Actually, do I really want this?" I am interviewing the company as much as they are interviewing me. I ask a question at the end to try to gauge a feeling of the culture of the place.'

Highlight relevant skills and experience

Once you find out what the company is looking for and what they need, highlight your relevant skills, contacts and interests. Emphasise your suitability to the production. Make it easy for them to hire you, tell them what value you could add to the production and what experience you could being that would benefit them.

DON'T

Forget your manners

Be courteous and polite, don't act inappropriately – a handshake is fine but kisses are too familiar. 'Don't be too informal,' says Julia Waring of RDF. 'On two occasions I've had boys giving me a kiss before I have even talked to them. I said to one of them, "Why did you feel it appropriate to kiss me, out of interest? I've never met you before?" He said, "My mum told me to do it." Of course he got never got the job. Behave appropriately. Take it seriously even if you are going in for a chat. People are extremely nervous but hopefully the person you are going to meet will take that into consideration.'

Don't appear rude or disinterested, even if it is a cover for being nervous; be engaging. 'Look someone in the eye,' says Julia Waring. 'I interviewed this lovely boy who I really wanted to give an internship to.

He had done media at uni and he had been managing a pub for a year so there were different sorts of qualities in him. He came in and he looked at the floor, and just occasionally he'd look at me. I told him that as hard as it was, he had to look me in the eye. The whole thing about being a researcher is that you have to have contact with people; you're not going to get anybody to talk to you if you can't look at them.'

Answer your mobile phone
Remember to switch your phone off before you go into an interview. You don't want it to go off during your chat and you definitely don't answer it. 'I've had that in interviews. And that's not just new entrants but producers – I think that's really rude,' says Jo Taylor.

Admit you're not prepared or don't watch their programmes
If you haven't bothered to watch their programmes, swat up on them or at least try to blag your way through the interview.

Ask about money
This is bad form and too pushy. You can discuss and negotiate your rate when they offer you the job.

Ask what the hours are
You'll be expected to work the hours the production demands – hinting or saying you don't want to work anti-social hours will not get you the job. It's pre-requisite at entry level.

Eat too much or too little before the interview
Eat too much beforehand and you're stuffed, uncomfortable, lethargic. Eat too little your stomach will rumble and feeling hungry might add to pre-interview anxiety.

Overstay your welcome
Let them close the interview; look for unspoken signs that it is at an end.

PROFILE
Kate Phillips, head of development, BBC Entertainment

What does your job involve on a day to day basis?

My job involves overseeing everyone who works in development in entertainment and inspiring the team to come up with ideas for all the channels: BBC 1, 2, 3 and 4, for all subjects and all slots, and to make sure that those ideas are slot driven, they are formatted and focused on the international market. Every two weeks we pitch to commissioners. We have a 50 per cent guarantee that 50 per cent of all shows commissioned come from in-house. Then 25 per cent come from indies and 25 per cent come from the WOCC (the Window of Creative Content). I now run all of the development teams so that we are all competing with the indies rather than each other and can react quicker with an idea.

How did you get your first break in TV?

When I was at Bristol University doing a politics degree I wrote to Esther Rantzen to ask for a job as a runner on *Hearts of Gold* and I got one! I moved to London and then I got a job working as a secretary. I got a big break when I was secretary at Prospect Pictures. I met Kevin Lygow (now head of Channel 4) who was an entertainment producer working at a company, Action Time, and he was offered a job at the BBC. I spent my lunch hour helping him with typing and things for Action Time as they were in the same building as us. When he was offered a job at the BBC he asked if would I like to come with him as a junior researcher.

Did you have a game plan?
No. I'd be amazed if somebody tells me they had a game plan and they actually stuck to it!

What's the best advice you've ever been given?
Indifference is the enemy, not contempt. And never buy cheap vegetables!

Why did you choose to focus on working in development?
I was always coming up with ideas. I used to write a lot of stories when I was younger, and I wrote my own stand up comedy material at university, so I think it was always there. But when you are producing you are a lot more restricted by budgets, dates and talent whereas in development it's much more a blank page. It's also how you come up with ideas, it's really conversations you have with people. Mark, Steve (her business partners in Mast Media) and I were talking once about how hard parents have to compete for school places and we came up with this idea *Beat 'Em to Eton* which turned into a parent competing for a fully paid public school place for a child. It came from that conversation. I really enjoy development but I do miss production as well.

What's the best thing about working in television?
Probably when you tell people what you do and they're really amazed. It's not like saying you're an accountant or an estate agent, it seems like a glamorous profession and of course it's not. But it is exciting when you're new to it. When you're doing a live weekly show, it's a real adrenalin rush, you're aiming for the show each week.

What achievement are you most proud of?
Professionally, running my own format company with Mark Baker and Steve Havers and paying our mortgages throughout! Personally, my three children!

What's the worst thing you ever had to do working in TV?
When I was a runner I had to go to the Courts of Justice for a chat
show to try to get access to someone who had been put in prison
for murdering his wife's parents. He was appealing and they were
trying to get him off. My producer made me to go into the court
during their lunch break and say to the girl and his parents, 'Hi,
I'm from a Channel 4 chat show, if he gets let off can you come on
our show?' It's just awful going up to someone in a situation like
that. It's like the newspaper world where you have to doorstep
someone.

What personal qualities have helped you?
I love watching TV and I really enjoy chasing the next commission.

What's the best way to get ahead in television?
Confidence and collaboration.

Chapter Four
First jobs in TV – running and researching

The runner

The first rung on the TV ladder is a runner. This is the first paid position in television. It's a good place to start if you don't want to be paid in hard cash but in hard earned experience – it's long hours for low pay. At the time of writing some TV production companies are offering £221 per week as a runner's wage, the minimum they can possibly offer, others slightly more.

Running is a good way to get a creative job in TV; you get the chance to learn a lot and you're in a good position to make contacts, get noticed and progress. Most successful TV producers, presenters, commissioners and people who own their own companies started as runners.

'Runners are called runners for a reason,' says Katie Rawcliffe. 'You literally do run yourself into the ground and do everything. I did. But I learnt a lot.'

What does a runner do?

As the name suggests, a runner is someone who runs errands, gets lunches, makes tea, collects and delivers tapes, and is a general dogsbody for people too busy to do those things for themselves.

Puja Verma's first job in TV was as office runner at Zig Zag Productions and on a three month contract. She says, 'I did anything that's asked – running errands to the post production house, getting my boss's breakfasts and coffee, delivering tapes, shopping for whatever's needed and translating anything that's Indian related.'

There are different types of runners. Office runners will help with general administration.

As a runner working on a specific programme you'll get the chance to go on location, help on shoots, and help to find contributors, locations, stories and facts and even shoot a camera.

'It's completely dependent on whether they are working in an office or on a production as to what they are expected to do,' says Julia Waring. 'Runners are there to pick up information and to be of service to the rest of the team. They need to use their initiative to make sure that shoots and other aspects of the production go smoothly. They have to be willing to do every job with a smile. They have to approach us intelligently and understand what their job is. As they are given clear instructions then they should be able to work to those instructions. It's all about clarity in every job.'

Helen Beaumont is a runner at Endemol on a Channel 4 series called *The Sex Education Show*, a varied role which has allowed her to help with research. She says, 'I work alternate days in the office and out on shoot. In the office I help out with the general admin: photocopying, faxing and helping out on the accounts. I work alongside the researchers reading newspapers and researching stories on the internet, I'll be set tasks such as finding locations and accessing props.'

Helen has been on many shoots and worked in the studio. She says, 'Out on shoot, I'll be doing everything and anything: looking after equipment, assisting sound and camera, looking after the presenter and crew, gathering release forms for locations and contributors, finding props and working alongside the director to allow the shoot to go as smoothly as possible.'

As budgets are squeezed there may not be enough money for a researcher on a production and so there are opportunities for fresh, bright and intelligent runners to rise to the challenge and do work that a researcher or assistant producer would normally do. If you can shoot a camera well, there's a chance that your material could end up being used in a finished programme.

Glen Barnard started running at BBC1. He says, 'I have done three different productions. In all of those I was doing a lot more than general running. They were virtually researching jobs as I was self-shooting things for edit and interviews. I did a lot of surveillance work, too, shooting under-cover camera.'

RDF uses a cover runner system to try out new entrants to see if they've got what it takes to work in TV. 'We invite people to come in as cover runners,' says Julia Waring, head of creative resources. 'They are people that we have met that we feel might have potential. Cover running is a very good way of seeing how people work within RDF, within a team, and how much initiative they use. We get constant feedback from our reception manager who tells us how they are doing.'

How you survive running is a good determinant of how you will progress and your career will shape up in TV. 'Running sorts the wheat from the chaff as the early years of work in TV are all hard slog and if you don't get an appetite for it then you should back out soon,' says Conrad Green.

How you behave, perform and how well you integrate with other people when you're running will have a bearing on how long you stay at a company and how well you progress. You need to be super keen, enthusiastic and open to anything that's asked of you, whatever it is.

Richard Hopkins says the key requirements of being a good runner are 'good organisational skills, a driving licence, a willingness to work really hard all hours, no snobbery about making people cups of tea and toast. And accepting that your job is to be trodden on by every single member of production and that they'll smile at you whilst treading on you.'

David Minchin started as a runner at Talkback Thames. He says, 'Never think any job is too big or too small for you. If someone says, wash my apple, you have to wash it; you may not want to but it's about doing everything and being keen and enthusiastic. It's very easy to roll your eyes and say you don't want to do photocopying but I think you have to enjoy everything you do and be seen to be enjoying what you're doing and not to worry about thinking things are beneath you. Everyone has to start somewhere.'

Starting at the bottom can seem demeaning as running can be quite a servile role. 'You do have to do a lot of crap when you're a runner,' says Kate Phillips. 'I did horrible jobs. I've had people swearing and screaming at me. It's not your fault but you're the nearest person. You have to be thick skinned. You hear stories of cover runners who have to go to a chemist and pick up the presenter's Viagra prescriptions and pick sultanas out of their muesli. But you just have to do any job that is asked of you in television, you can't have an attitude. You do whatever it takes.'

Do what is asked of you, bide your time and remember that it's not for long. Think of it as just the first steps of a longer journey. Running is just an initial apprenticeship role that you'll be doing for a brief period and that leads to greater things. 'Anybody who is any good won't be a runner for very long. They'll move on,' says Daisy Goodwin. 'If I found someone who was very smart I wouldn't necessarily make them make tea but I'd make them run for a few months so they can learn the ropes.'

'You pay your dues at the beginning but the rewards do come,' says assistant producer Jenny Popplewell who got her first TV job as a runner on a five show called *That's So Last Week*. As she was quick to learn and hard working she was quickly promoted to junior researcher on the programme. The company then kept her on as a full time researcher and she stayed with them for three years rising to AP before moving onto bigger and better things.

Being a runner enables you to observe how the industry works. You're a good position to see what's going on without the responsibility of having to make it all happen. It's a great place to make contacts, learn as much as you can and progress.

'Once you have your first running job, assuming you've done well, people like you and appreciate what you've done, you get strong references and the people within that team will recommend you for other projects,' says Glen Barnard.

Running was a good starting point for Conrad Green. 'I answered the phones, made tea and generally did any crap job necessary for almost a year. Crucially, I also got to know people on all the productions who said they'd like to work with me. I think I was pretty good at it, but I remember that after about six months I got really frustrated. They actually offered me a job as tape librarian which would have been a quick promotion but totally down the wrong path so I stayed running.

'The breakthrough for me was when I realised I had to let them know I had a brain and could work in production, so I started writing a satirical work newsletter making fun of the staff in the style of a Private Eye piece which, looking back, was probably pretty embarrassing but it let people see that I could write, and made them see me in a different way. From there I got a job as junior researcher and I was off!'

As a runner you'll get to see inside a TV studio, work on location and inside an edit. You can watch others, see what they do in their jobs and who gets on and why without the pressure of being a researcher or producer. Running is a great place to experience the different aspects of TV production and work on your networking and team working skills.

Helen Beaumont has learned a great deal since she started running in 2005. 'I've had lots of brilliant experiences. Working on *Gladiators* was a huge amount of fun. I had to try out the games, so that the cameramen could work out their shots. Very few people can boast they had to play a game of Duel, run up the Travelator and climb up giant sponge pyramids because it was part of their job description! I had a great time looking after the dragons on *Dragons' Den* and I've had a fantastic time working on *Sex Education*, it's all incredibly interesting.'

David Minchin loved the variety of his job on *X Factor*. 'It was a brilliant first runner's job as there are so many parts. I was a producer/director on the audition part as it's basically split up into different crews. While I was doing that, I had to get release forms and log the good bits of filming as the director hasn't got time so you have to be their eye and ears. When they're editing, they have 5,000 hours of footage for a one hour show, so it's good for you to see what are good emotional moments and one liners.'

As well as learning and contributing to the material of the show, David had to look after the contributors. 'People management,' he says. 'Getting contestants from one room to another on time and dealing with hysterical parents who think their little darling is the best thing since sliced bread! It was really good but hard work with long hours. With auditions you're away for seven days at a time and working 15 hour days.'

And once the show moved out of its audition phase into transmission the pressure and the hours increased and there were new demands. 'The studio side was different. It's less intense as you're in the office in the week, helping out and maybe finding time codes and footage for VTs and helping producers,' he says. 'On studio days I was in charge of looking after the judges so I got to take tea to Simon, Louis and Sharon which was really good. They were very nice but they are very busy people and sometimes they just don't have time and can be very abrupt. You have to deal with it and keep smiling. But it was a great first job.'

Runner Glen Barnard finds his job fascinating. 'Every time I have a tough week and I think I'm not getting paid enough, I speak to my friends and they say that what I'm doing is amazing. Every day and every week is different. I love that.'

Working as a runner is a good way of getting noticed. Never miss an opportunity to help or offer your services but don't be pushy. Use your judgement – if you ask at the wrong time you're going to annoy people and they won't want to work with you. There's a fine line between pestering too much and taking the initiative.

I've got a degree – do I still have to start as a runner?

The short answer is yes. Unless you're incredibly lucky, know the right people, have a family member already working in TV or have specialised knowledge or a skill that makes you indispensable, most people will expect you to start as a runner, irrespective of your degree.

'You kind of have all these dreams about coming out of uni and jumping straight into a job, but it took me a long time to find one although I had networked quite well when I was at college,' says entertainment researcher John Adams. He did work experience whilst still doing his BA in Broadcast Production at Ravensbourne College of Design and Communication. 'I didn't have the normal kind of uni experience like a lot of people. It was hard work schmoozing with as many people as I could. When I came out I tried really hard and I wrote to everyone. I spent all day looking for work. It took a couple of months before anyone really phoned.'

Running is at the bottom of the TV food chain and it can be boring and demoralising. You might have to do it for up to a year before you get interesting work, no matter what your degree, but it is a worthwhile experience to start as a runner and an essential rite of passage. Running is a good training ground to see whether you are right for the industry and whether it's right for you. 'When someone's a runner you can often work out whether they are the right person and if they have the right attitude,' says Tim Hincks, chairman of Endemol UK.

What do I need to do to get noticed?
TV is all about being on the ball. You always need to be one step ahead of what is going on. Plan – anticipate what's going to happen next and work out what's required and where you can help. On shoots or in the office think about what needs to be done and how you offer your services to help the production team.

'Work damned hard, don't be too cocky and always try and think of what you would need if you were in your boss's position and be prepared to give that to them before they ask,' says Conrad Green.

Using your common sense and working hard to help others is a good way of getting noticed. Observe what needs doing and do it without being asked. 'I look for somebody who goes that extra mile,' says Ruth Wrigley. 'If somebody in my position happens to notice a runner, then that runner has to be amazing. And that's happened to me once with a runner called Gus who's now a production manager; he was just brilliant. He'd enter the room and if there was nothing for him to do he'd go around and ask everyone what they wanted to drink, empty the bins. He never stopped, he was constantly asking, "What can I do to help?" '

'I put myself in the shoes of the producer and think, "What would I want from this person?" ' says runner Glen Barnard.

You have to think about what's required for the show, what the team are working on and where you can be of most use. 'I like to keep on top of things and think ahead,' adds Glen Barnard. 'If you go on a shoot and it's really hot, before people tell you they're thirsty, get water on your way to the shoot, be prepared. People will remember that. They are simple rules to go by, but go a long way.'

As well as thinking of others, you need to be unselfish and put the needs of the production above your own. The key thing that you need to be a good runner is to be one hundred per cent flexible in terms of time. You need to say you're flexible and can work all hours. Even if you are getting paid the minimum wage you're still expected to work long hours and be there while the programme is being made and there are people in the office. Being the first to arrive and last to leave always makes you look dedicated; being happy to do so will help enormously, too.

'Always have a smile on your face. You don't have to be hugely skilled to run, but to be part of a team you need to be an outgoing, approachable and quick thinking,' says Glen Barnard.

'I've always done everything I can to help,' says David Minchin. 'I've never said, "Come on, it's late, let's go home. I'm getting bored." As a runner, think about the next thing your director is going to ask you; really obvious things like work out when your tape is going to run out and have a new one ready. It's all about keeping one step ahead of the game and anticipating what is going to happen. People are often impressed when they say, "Can you call John and say we're going to be late" and you say, "'Oh, I've already done it because I noticed we were late." '

Coming up with ideas and facts, locations and contributors for the programme is another good way of getting noticed. Make sure you get involved and become an indispensable part of the production team. TV is very egalitarian – you'll have the same opportunities and be in the same meetings as APs and researchers, so if you have good ideas make sure your voice is heard.

On the entertainment shows I series produced, the runners were a key part of the teams making the programme. I always gave the runners on my shows the chance to get involved in the brainstorms where we came up with ideas, games and set ups for the shows. I was always open to hearing their ideas and thoughts and welcomed their input, even more so later when they went on shoots. As they had been on location and seen the shows being made they were in a good position to make judgements and offer suggestions to improve the programme. Good ideas are always welcome and are a good way of getting noticed especially from newcomers who have fresh new ideas and a different perspective on things.

How do I go from runner to researcher?

Once you have experience as a runner you can decide which path to take: either to work in creative TV production in editorial and be at the mercy of the freelance marketplace or in production management (the money/logistical side) which tends to be more secure in terms of longer contracts with more stability.

If you want to make, create and produce programmes and work in editorial the next rung on the job ladder is a junior researcher. Working as a junior researcher is a good way of learning the ropes and testing whether someone can cut it without being given the full responsibility of the role. A junior researcher straddles the runner/ researcher role and will be involved in doing both.

It is a mixture of research and running. As money is tight, and in these times of lean budgets, you will probably be a researcher in all but name, working under a junior researcher title but without the experience and be required to do the job of a researcher and for much less money! In some cases you might even be doing things that an assistant producer would do – if you are the only researcher working in a small team with just a producer.

Researchers work on a programme-to-programme basis and are on short-term contracts, depending on the duration and the run of the programme. I've had six-week, three-, six- and 12-month contracts as a researcher. They vary depending on the type of show you're making. On a documentary you might be the only researcher working with the producer/director and possibly a runner. On a large entertainment show you might find yourself one of several.

To be a researcher you must have experience in television as a runner, have proven skills and experience which are transferable or have specialist knowledge, languages or unique skills and contacts that the production needs.

The main ways of getting promoted from a runner to researcher are:
- Being noticed by your last producer/series producer and promoted on their next project
- Advertising yourself on a website
- Applying for jobs on websites
- Word of mouth and recommendation

The last of these options is the most likely route of progression. People who know you and trust you are more likely to promote you than someone who doesn't. If you have impressed your executive producer, series producer and production manager they are more likely to give you a break as you are a known quantity. They'll like you and want to work with you. They'll know that you've more than fulfilled your role as a runner and have done research and will know what you are capable of.

If they don't know, make them aware in the nicest possible way. Write them a thank you letter for giving you the chance to learn so much; make sure they see the briefs you've written, the locations you've found, the contributors you've nurtured. Make sure they know because that crucial first credit as a researcher is important, especially at the beginning of your career.

If you find that as a runner you are doing the work of a researcher do ask if you can have a researcher or at least a junior researcher credit. Forget asking for a pay rise. You can try but you'll be in the budget as a runner and will be in a specific pay band. But you can get your payment in kind by being recognised for the role you performed as researcher. You can put it on your CV and they will confirm that. Getting your first researcher credit will help you get your next job in that role.

Most producers will be happy to do so as they'll recognise your contribution to the show and will be happy to return the favour. When I was at the BBC the graduate trainees in the department brainstormed games, rounds and ideas for a series I produced. I recognised that they played a crucial role in helping me to develop the format and I made sure each of them had a junior researcher credit on the closing titles which they could put on their CVs.

If someone doesn't agree don't put it on your CV. If they check and your last employer said you were a runner, it won't reflect well on you. List yourself as runner/junior researcher and list the different tasks you performed and the things you achieved for the programme.

In most cases your series producer will recognise your talents and if they can they'll take you with them to their next production. If I thought a graduate trainee or runner had potential and had done a good job I would offer them a job on my next production – promoting them to junior researcher or researcher – giving them a break and helping their career because I thought they could deliver. A series producer wants to work with good people who they can rely on to be a good team member, who works hard, comes up with the goods and contributes creatively. I would always favour someone with these attributes rather than a lazy, burnt out AP with more experience but less talent.

PROFILE

Daisy Goodwin, managing director, Silver River Productions

Starting as an arts producer at the BBC, Daisy devised the highly successful shows *Bookworm, The Nation's Favourite Poems, Looking Good* and *Home Front*. In 1998 she moved to Talkback Productions as head of factual, and was promoted to editorial director. The three BAFTA-nominated shows *House Doctor, Jamie's Kitchen* and *Grand Designs* as well as *How Clean Is Your House?, Would Like to Meet, Your Money Or Your Life, Property Ladder* and *The Sex Inspectors* were all devised and brought to the small screen by Daisy.

How Clean Is Your House? is one of FremantleMedia's most travelled factual entertainment series. It has been sold in 16 territories around the world including Australia, Malaysia, Thailand and the Ukraine.

Daisy is a mother of two and editor of numerous poetry anthologies, including the bestseller *101 Poems That Could Save Your Life*. Daisy has also presented the BBC2 production of *Essential Poems (To Fall In Love With)* followed by *Essential Byron, Essential Poems for Britain* and *Essential Poems for Christmas*. Her most recent project is *Reader, I Married Him*, a series about romantic fiction for BBC4.

What does your job involve on a day to day basis?
I'm the major shareholder of Silver River, I own the company and I'm responsible for getting business, overseeing production, quality control, having ideas and managing my staff.

How did you get your first break?
I had a very good CV because I'd been to Cambridge. I was very persuasive, I had lots of things to say, I was incredibly confident. I went to film school in the States for a couple of years, then I did the BBC associate producer's training scheme for two years.

What qualities have helped you succeed?
I think I'm reasonably intelligent, I'm quite good at selling. I've got a fairly broad frame of reference, I enjoy doing the things I do. I like taking risks, creatively. I'm passionate about what I do.

Did you have a game plan?
When I started I wanted to be a director. I just wanted to make films. At some point you have to decide whether you want to be a producer or a commissioning editor. I decided producing was more lucrative.

What advice would you give someone starting out?
My advice to anybody is the riskier the better, the higher the risk the greater the rewards.

What's the best thing about working in television?
Well, I hope it's yet to come! Having a hit show is very good news, it's great. Winning awards is great, selling a show is always good, getting a commission is exciting. Those are all the things that keep you going. It's not really about the money. Affirmation is always exciting.

What's the worst thing about working in TV?
There are days when you just think it's going to be a disaster. You have to constantly remind yourself that it is only television. Nobody died.

You have had a prolific hit rate of TV hits – how have you done that?
Every two or three years it's time to think about what can I come up with next? How can I do something different? It's very easy to get stuck in a rut. You've got to keep moving on. It's very important to become very good at something, but then to be thinking what can I do next? Because no one is going to re-invent you, you have to do that.

Do you think you'll still be working in TV in 10 years time?
I've no idea, but I'd probably want to be. I hope somebody else will be running Silver River full time and I'll be writing books.

How have you found the time to write books as well?
Everybody says that! It depends how much you want to do things.

How do you get a job in television?
If you want to do something and you're energetic, creative and adventurous then you'll do very well. If you're an entrepreneur you'll probably find a way of doing it. I'm a big believer in helping young people and giving them breaks but nobody needs to work in telly, if your good at it, then you'll do well. I don't think there are very many very talented people who aren't doing well. It never hurts if you've worked on a hit show. But you never know when that's going to be . . . The moment you start to whinge, you shouldn't work in television. Nobody likes a whinger.

What is a researcher?
Being a researcher is the toughest job in TV. Without a researcher finding contributors, locations, stories and creating storylines there *is* no show. You can have the best show idea but you need the right contributors to make it stand up and happen.

A TV researcher finds contributors, facts, storylines, locations, ideas and might even shoot a camera, if they are skilled enough. They'll interview, meet and vet potential contributors, assess them and produce them for the programme. They'll be the first port of call for a contributor but will also help set up and go on shoots on location and in the studio. They'll support the edit finding additional information, clips, stills and even shooting extra interviews and material.

Researchers tend to be given tasks to get on with (though sometimes they might not even given a clear brief but left to work it out themselves). You need to have a lot of initiative and self reliance. Good researchers are methodical, organised, and excellent at finding things at short notice and under pressure.

You have to be flexible and think laterally; you need to be able to come up with ideas quickly, have a good knowledge of what's going on in the news, who's who in film, TV and celebrity. You might have to operate a camera on a shoot, find a location or facts for a busy producer/director, book a celebrity or track down an unwilling contributor and persuade them to go on camera. The producer wants results not excuses and you have to be able to deliver, often without thanks and sometimes to a different brief from the original as things might change. You may have spent painful hours finding and securing that crucial contributor to then be suddenly told to drop them. It happens all the time. There's nothing worse than working on a live show and an item being dropped because the show is running over. You are there with the contributor as it's happening and you have to break the bad news to them.

It's a regular occurrence because items run over their allotted time slot and so others have to be dropped to make sure the show hits the ad breaks and the off air time. Presenter Anna Richardson got her first job in TV as a researcher on Channel 4's live morning show *The Big Breakfast*. 'I worked on the "Family of the Week" item where an ordinary family had to stay at the house for a week and be part of the show. I was told that the dad of the family loved Potter's Pork Pies. I had to ring up the Potter family in Yorkshire and explain the situation. I told him that we were setting the dad a question on the show with the chance of him winning a lifetime supply of pork pies and would they do it. They said yes and overnight they had to make 500 pies and get them shipped to London by 5am! At 5am that morning I turned up and all the pies were there. But the item was then dropped at the last minute. I had to apologise. He went mad. I had to say, "Sorry, but that's live TV." I was told to say that but I felt humiliated and so awful for him, he was saying how it cost them so much money to make all those pies.'

Researching is incredibly specialised and it's knowing how and where to look. In the past, people worked as researchers for their entire careers, staying in that one role for years at a time, building contacts and knowledge. When I worked at LWT (now ITV Productions) I worked with researchers in their 30s and 40s. Now young people in the industry think of researching as a stepping stone to producing and directing and want to do it briefly before moving on. It is, however, an important and crucial job in its own right and it should be mastered before you think of moving on to the next level of assistant producer.

What does a researcher do?

Using the same skills as a journalist, a TV researcher is required to become an expert on a particular area, topic, guest or subject relevant to the programme. They need to be able to understand and disseminate relevant information clearly and write a brief, script or notes for a producer or presenter that can be easily digested with all the salient points to the fore. As well as uncovering information and writing, TV researching has many different requirements.

'Being a researcher is a varied role,' says assistant producer Harjeet Chhokar who has researched on *Holiday Showdown*, *First Edition* and *Liquid Assets*. 'You can be required to do anything from logging, people finding, fact checking, expert casting, writing treatments, setting up locations, recces, investigating stories. That is one of the great things about being a researcher, your job is really varied and interesting.'

A good researcher finds out the facts, statistics, information and the experts who hold the key to an area of interest quickly, under pressure and with ease. They always deliver, never complain and always answer yes – whatever the question! Even if the goalposts are continually moving you need to be calm and tenacious.

A researcher's job varies from programme to programme depending on the genre, type and shape of the show. No two research jobs are ever the same. On a reality show like *Big Brother* you'll be finding and casting contestants; on *This Morning* finding stories and experts for live items; on a documentary finding locations, contributors and managing story lines.

'It all depends on the type of programme you are working on,' says Harjeet. On something like *X Ray*, the consumer programme I worked on, I was checking out potential stories and working them up to see if they would be suitable for that week's programme, whereas when I worked on *Real Story* with Fiona Bruce, it involved working on just one story and doing everything from people finding to recces to fact checking. On *Holiday Showdown*, I was casting, setting up locations, and once out filming, was shooting on the Z1. Development was different – I was heavily involved with brainstorms and then writing treatments and pitches.'

Each programme I worked on as a researcher had different needs, a different brief and was in a different genre. Each show has unique demands and you need to start afresh each time. I never felt that even though I had been a researcher for a few years that I could relax as each new job presented new challenges and I'd have to work out how to best meet them.

'Things are never the same on each production,' says Simon Warrington, an AP who worked as an entertainment researcher for five years. 'It depends on what show I'm working on but mainly my job involves writing biographies of the guests due to appear on each show, choosing clips and editing the insert role for a live show and sourcing extremely difficult guests, locations and products within an unrealistic time-frame!'

'A researcher might have to do a huge amount of unnecessary research into subjects that your producer thinks might or might not be necessary for a subject,' says Richard Hopkins. 'When you're whittling down the final contents of a show, you cast your net quite wide so you'll end up researching many different areas that end up not being used in the show. But if you research them well you'll find you'll be incredibly important to the show.'

What skills do I need to be a researcher?
Even though the nature of researching changes from show to show there are some common denominators that TV researchers share regardless of the genre. Ultimately, they're the same skills you'll use and rely on throughout your career in TV and will become part of your make up. The essential skills are tenacity, an ability to work hard, resilience, optimism, being organised, methodical, and creative thinking outside the box.

'The skills needed to be an effective researcher are an enquiring mind, perseverance, a willingness to work hard, people skills, empathy and some salesmanship,' says Matt Born of DV Talent.

A good researcher never gives up. Television is full of rejection and disappointment. It's important to be able to deal with rejection when working in TV because you'll be faced with it while in and out of work and you'll face rejection when you're trying to find contributors and persuade them to appear on your programme. Bouncing back unscathed with unrelenting, undentable enthusiasm will give you the edge and enable you to survive in TV.

You might shake in your boots when you're asked to deliver the impossible, I know I used to. I'd agree and put on a brave face; I'd go away and try to figure out the best way to find what was required – be it a location, a particular type of contributor, archive, prop, idea or fact. Go away, work until you find what you need and go for it; work until you find it. Then when you deliver, the producer will move on to another seemingly impossible demand and moved the goalposts – but that's telly!

If you can't find out what you need, you must find the solution for a good alternative – one that's even better than the original idea – and then be able to sell it to your boss who might be wedded to his original thought. You need to be a good lateral thinker. Even if the answer's not immediately obvious, you can find things by approaching the problem in different ways. TV is an endless source of conundrums which you have to solve. You just have to be prepared to meet the challenge and keep your wits about you!

You can't be expected to know everything, so don't pretend you do – ask someone who can help. Remember, you are part of a team and can call on the resources, knowledge and experience of those around you. Similarly, you should not be afraid to share information and help others. Believe in karma – what goes around comes around. If you help others they'll be more likely to help you in the future. If you have good contacts or have ploughed a furrow that might be of use to someone else, don't be afraid to share it. Be generous. People will be more willing to help you out when you call on them.

Having a strong journalistic flair helps. Knowing how and where to look and find stories, ideas and characters is essential. It's about knowing who and how to ask.

'You should have a journalistic ear,' says Richard Drew. 'Phone conversations should be like chats with friends. You need to be able to get contributors to open up to you and to ask the right questions so you get all the juicy stuff from them. You have to be able to dig deep.'

Make sure you use a variety of sources in your quest for information. Don't just rely on the internet. Julia Waring selects researchers by how good they are at thinking laterally and how resourceful they are. 'They will naturally be the sort of people not to rely totally on the internet. They will use it as a tool rather than a means to an end. You can't just say "I am not working today because the internet doesn't work." '

You can use the web for research but thorough in-depth research is a lot more than that. Use other sources that will give you good leads such as newspaper articles, academic studies, opinion polls, specialist magazines, industry survey statistics, Government reports and books. You can use them to find experts and then contact them. Get on the phone and talk to the people who know. They'll be able to help you find the facts and give you informed opinions.

'The number one skill you need is phone bashing and unfortunately it's becoming a lost art in this internet age and it's a pet peeve of many series producers,' says Richard Drew. 'You have to be able to pick up the phone, charm people, and get them to do your show.'

You need to get the best experts and advice as TV is competitive – you want your show to be compelling and unmissable television, and that's down to finding and producing the right stories with the right contributors.

'Whether it's a contestant coming on to your show, an expert giving an opinion or a salient piece of information, you've got to fight to make that as brilliant and as thrilling as it can be, even if it's only seven seconds of information,' says Richard Hopkins. 'It's got to be better than any other seven seconds of information you might have. The more short cuts you take, the more drab the show will be, and the worse you'll be remembered as a researcher.'

As well as being a tenacious people finder and a creative storyteller a researcher needs to be organised and meticulous in their approach. 'I tell all my researchers to keep a table of all the people they've contacted, their responses, who they're waiting to hear from,' says Richard Drew. 'It helps me keep track of how hard they're working and is also useful for future series and their fellow team members, so you don't end up calling the same people over and over again!'

Make sure your desk and your computer desktop are clear, and your notes easily accessible and well filed – you might need to find something at a moment's notice and you need to have systems to make that easy.

'The organisation of a researcher is crucial. Your producer needs a bit of head space to come up with the creative ideas and you need to run around them making sure that they are being serviced for all those ideas and that all the props are there for the ideas, that the filming schedule is thoroughly understood by you and the rest of the team. So you're constantly working around their potential weaknesses. If they're a brilliant producer you'll enjoy it even more,' says Richard Hopkins.

It's universally true that no matter what the genre, whether factual or entertainment, you need good journalistic skills. You have to be able to write well, find facts quickly, track down experts and contributors, interview them well, find locations and props. You might even be a dedicated archive researcher tracking down and clearing archive material – old film clips, adverts, TV programmes or material from someone's personal collection.

You also need to be creative in whatever genre you're working in, from coming up with ideas for games, gags and items on an entertainment show to storylines, experts and managing problems on a factual show.

Some researchers like to specialise in becoming good people finders, contestant researchers or booking celebrities. Some become shooting researchers who film on location; others like to specialise in development, coming up with ideas for programmes.

Whatever the genre of researcher, you need to be persuasive and persistent. It's a tough job, finding what you need is not always easy and when you find it, people are not always co-operative or willing to do what you need them to. You need to be positive. You need to be precise.

'You need to keep going until you've got what you want, really want, without breaking the law,' says researcher John Adams. 'You need to be patient, although I'm terribly impatient, as sometimes it takes a long time to find things but a lot of the time it is out there.'

Tenacity and lateral thinking are key. Instead of giving up you have to think of new ways to find what you need. If something doesn't work, think again.

How do I research?
Former TV series producer and media lecturer Lucy Reese suggests these basic golden rules for research.

Golden rules for researching

1. Talk to people on the phone
2. Read newspapers – local journalists are a mine of information
3. Consult reference books – phone books, electoral register are useful research tools
4. Talk to organisations – schools, libraries, community centres. There is a national association for almost everything you can imagine
5. Walking the streets and hanging out in areas where you think you could find the right people

With the internet it's easy to research your subject – whether a guest, a topic or a contributor. Most companies subscribe to a cuttings service like Lexis Nexis, an archive of content from national and local newspapers, magazines and legal documents all on one site.

Most companies usually subscribe to the *Red Pages*, which gives you the contact details for TV and film stars, celebrities and sportsmen. You can use it to track down the contact details for a famous face in seconds. You can use relevant websites to post advertisements and search them for information and contributors.

The internet is an invaluable tool for celebrity information. Gossip pages like Perez Hilton and Female First as well as a celebrity's own blog pages are a useful first port of call as they often break a story before the papers. They are very entertaining and can also be used for brainstorming new ideas. Says development AP Jenny Popplewell, 'It's important to get a whiff of a story before the papers because you can bet every commissioner will have the same idea when a good story breaks.'

Use newspaper articles to gain facts. Speak to the journalists who have written the articles and track down the experts and case studies they've used. Go to the library and use reference books, the electoral register and the phone book to find people quoted in newspaper articles.

Talking to people is a good source of ideas and stories. 'I have an in-built story radar which is very rarely switched off,' says freelance print journalist Rachel Roberts. 'So I scan local papers for quirky tales and watch local news. I listen to friends' stories about what's happening to other friends, read as many newspapers and supplements as I can to spark new ideas. It's a reflex action to turn any interesting information into a possible story.'

Finding contributors

When it comes down to it there is no substitute for phoning people, making contacts and exploring all avenues within a genre. There are organisations for everything – charities, associations, community centres and groups. Start at the top and branch out finding out who's who. Find the key players in an area, the people who run the associations and community centres, and get in touch with them and speak to their members, win their trust and unlock their contact books and their minds. Speak to everyone you can. Tell them what you need and gain their trust and help in finding it.

Once you've made contact with them by phone meet them face to face so they know who you are and you can win them over. Spend time chatting to them, listen to what they have to say (sometimes not always relevant) before moving on to your own agenda to uncover what you want. Give them time to come back to you with ideas, names and contacts.

Be methodical in planning your approach and keep proper records of who you have spoken to and when. Put together a hit list listing names, position, contact numbers and what you've agreed and who they know. Create a file on that person or group, take notes when you meet them and type them up into a brief. Make sure you get names (correctly spelt) and telephone numbers as well as email and internet addresses. Leave no stone unturned, your research should be exhaustive so that you know all there is to know about a subject and all the people who work in it.

Speak to people in associations to find out who they know in their organisations who might be able to help, who have value or are good characters. If they're not right and can't help find out who they think can – get names and numbers and these will help you find the information you're looking for. Once you have made contacts, won their trust and started putting the word out, leads will lead to new leads.

'Because even if they're not right or can't help you, they can often point you in the right direction, save you time going down wrong avenues, refer you to a friend, or help you look. If you can phone bash you're instantly miles ahead of your competition,' says Richard Drew.

Use associations and clubs to find characterful contributors. Find the academics and experts who populate an area. Phone them as well as email them, arrange to meet them if you have time – even if at first they are resistant, be politely persistent. Sometimes talking to people can save you days of fruitless internet research. It may appear more time consuming at first but if you take time at the beginning of a project to call and meet the right people they will come back to you with leads, contributors, facts and what you need.

When you need to find experts, contestants or contributors to fit a brief for a documentary you have to hit the phones. Phone bash, phone bash, phone bash! You can post ads on websites, company intranets, place newspaper adverts, put up posters and distribute flyers but the key to finding people is to pick up the phone and talk to them and then go and meet them.

Listen to everyone and talk to anyone who could help you. They might not be right but they'll know people in their field and might be able to find people that can. It's like six degrees of separation. You just have to ask enough people to find what you need. It can be tough. Often you will be looking for a specific type of person to fit a specific brief which can sometimes seem an impossible and daunting task.

'Figure out what type of things possible contributors might be interested in, find them that way through events or organisations,' says shooting producer/director Jon Crisp who specialises in making documentaries. 'Then use your instincts to decide whether they are good enough characters and have strong enough stories to make them right for the programme.'

Sometimes, however, you might feel that you're hitting a brick wall and coming up with nothing. It's good to try to distance yourself and review your work. 'Think laterally,' suggests Richard Drew. Go back to the beginning. 'When you've exhausted the obvious routes, take a few steps back and brainstorm other ways of finding contributors. Don't just slog on relentlessly. You have to be able to assess what's working and what isn't.'

Actually hitting the streets is a good way of finding contributors or experts; it can be time consuming but it can also be worth it. You can assess and find a contributor instantly – nothing beats meeting face to face. I've found many people in this way for TV shows and adverts.

It's not always easy to find the right person and it pays to keep calm and keep going. 'It's very hard if you're asked to do something and it doesn't exist,' says John Adams. 'You can sometimes be in big demand. If you are doing a fast turn around live show, and everyone is very stressed, it filters through and the researcher can sometimes be near the bottom of the pack. But you've got to learn to take it on the chin.'

You face the challenge of looking for contributors with all manner of unique and unusual predilections. For example, people who like having sex with strangers through the internet or who marry mail order brides or mothers who wet nurse other people's babies.

'Before I got a job on Alan Carr's *Celebrity Ding Dong* I was writing a show called Granny Fanny,' says researcher John Adams. 'It was a documentary series following the sex trade in old people. There are websites where people list the grannies they'd like to have sex with. I was thinking, "Why am I doing this?" When I was a kid I wanted to work in TV because I watched *Noel's House Party*! I didn't think I'd be working on Granny Fanny!'

Dominic Crofts got his first TV job as a researcher on daytime chat show *Kilroy*, a notoriously difficult but good training ground. 'That was a really, really hard job,' he says. 'The producers would tell you what the subject was. It could be a neighbours from hell or my brother sleeps with my sister sort of thing, and then you had to find people that had. They had a big database from which you could call people but other than that it was phone bashing, going on the internet, going through newspapers, and calling up estates and seeing if anyone had had a row.'

David Minchin spread the net wide to find someone to take part in a pilot show called *The Boss Project*. 'We made a poster and sent it out to the right sorts of areas, bingo halls, libraries, public swimming baths, which got a few results so we changed tactics and contacted the British Chamber of Commerce which works with small businesses. We had to be very charming. We asked them to send out this poster to their members and although most said no sometimes you get lucky. We found one person through 15,000 emails!'

'It was quite difficult to find contributors because, one, getting people to nominate their boss and say he is rubbish is hard enough and two, once they've done that, convincing the boss to let you film them and say they're not very good and they could be better is even harder!'

Often you'll spend fruitless hours tracking down the right contributor and then find they might not want to take part. You have to be prepared to keep approaching the same people again and again not taking no for an answer. You also have to make sure you get there first so you get access to them before everyone else.

Use your judgement to find the right people and to make sure they are reliable and credible. Make sure they are who they say they are and with evidence to back up their claims. Documentaries in particular rely on the strength of their characters and poor research can end lead to them being scrapped if the contributors are weak. Something that no one wants to happen.

'With contributors, character is key,' says David Minchin. 'There's nothing better than a character that you hate and love at the same time. You need someone you know will go through with it because obviously you don't want them pulling out half way through the show.'

A good way of telling whether someone will be any good is to put them on camera. Jon Crisp says, 'You can also tell with experience, when you meet people, how passionate they are about the subject matter. You always want to find someone who is keen to talk about their story rather than just has a desire to be on telly.'

Always make sure you write down who you've spoken to and when and what was agreed. You will end up speaking to hundreds of people so your notes should be clear so you can refer to them easily or if you need hand them over to your AP or producer to read.

Never throw away a number. Always maintain good relationships with your contributors, contacts, journalists, press officers, people who run associations, clubs and groups. You'll then be able to call on them time and time again.

Casting contributors and experts is key to so many different genres of programme whether a topical chat show, game show or a documentary, and it's the researcher's job to find them, judge them and 'sell' them to the producer. You have to imagine how they would look, sound and appear on TV and evaluate whether you would watch them. Are they:

- Who they say they are?
- Knowledgeable and informed and can they deliver this experience on camera?
- Natural on camera?
- A good character?
- Credible?
- Compelling and interesting to watch?
- Attractive, good looking and easy on the eye?
- Funny or quirky?
- Do they have memorable character traits?
- Unique and unusual in their views?
- Do they have a USP – a unique selling point?

You can see the see why Cheryl Cole, Simon Cowell, Charlotte Church, Kerry Katona and Jordan are compelling to watch and why they attract massive audiences and have hit shows. It's using your own instincts and judgement to work out how much potential someone might have. 'In ninety per cent of the cases you can tell but then you have to be careful,' says David Minchin. 'There's such a difference between talking to someone face to face and then putting them on camera.'

Assessing contributors by phone

Because of time constraints, tight schedules and budgets it's important to be able to access and assess potential contributors by phone. Most casting researchers do it in this way in the first instance. While working at RDF I would hear large teams of researchers and APs phone bashing daily to find, persuade, engage and assess contributors for *Wife Swap* and *Holiday Showdown* before they met them.

'They have got to do that on something like *Wife Swap*,' says RDF's Julia Waring. 'You cannot rely on people writing in, wanting to be on a programme. Researchers have to use all those old skills, to be on the phone, all the time.'

Researchers and APs do go out onto the street and meet people on recces, but these are expensive in terms of time and money. The person has to be thoroughly vetted on the phone first to make sure they are absolutely right. Researchers have to work out if someone is worth seeing before they take the time to go and visit them.

When interviewing someone on the phone, ask yourself are they:
• Interesting and entertaining?
• Sane?
• Credible?
• Coherent and articulate?
• Knowledgeable about their subject?
• Able to withstand the stress and demands of filming without becoming camera shy?

Rigorously check someone out by chatting with them on the phone – discussing a variety of topics, gently digging to get as much background as you can before meeting them.

Do it with utmost meticulousness and charm so it feel like a chat rather than a grilling! Make sure that they know their stuff, are confident, amusing, and fully informed and briefed.

How to deal with and manage difficult contributors

As a researcher you are the production's, the production company's and broadcaster's first port of call with a contributor. Often you will have found them, nurtured them and built up a relationship of trust. As a researcher you are responsible for managing the relationship, keeping them onside and making sure they deliver on camera.

It can be a difficult relationship to manage as you'll be balancing their expectations and those of your producer who may want you to get them to do and say things that they might not want or agree to. Filming delves into all aspects of people's lives and exposes it for all to see; you have to build a relationship of trust and openness so people will let you in.

Many members of the public are now incredibly media savvy and might not let their barriers down. You have to work out ways of overcoming this so they are natural and candid on TV. You'll become the contributor's best friend for the duration of filming and it's your job to keep them happy and to make sure they continue to do the programme. They may rely on you for emotional support and guidance, sometimes at all hours of the day. It's one of the most difficult and challenging aspects of the job.

'If you have a contributor who turns on you and the team, the whole programme hangs in the balance,' says Harjeet Chhokar. 'This is why you must be a good people person, so you know how to deal with them, especially when they are upset.'

Dominic Crofts worked as a researcher on *Big Brother*. He says, 'I was on the care team where you had to find out all the things about the housemates who were going in the house in advance. You supposedly had to look after them and make them feel at home.'

He was also responsible for looking after their nearest and dearest too. 'I had to deal with their friends and family. That was a weird job.

When someone's relative goes on *Big Brother* and they do something on the show that might not be seen as good from their family's point of view, their family tends to get really angry and say the editor has totally changed the way it was. He didn't. You have to take the family member and deal with them as basically you're their point of contact throughout the whole run. You're always talking to them and making sure they're alright. It's OK when they win because they're really happy. But every other contestant hates you!'

It is important, therefore, to build a relationship based on trust and respect. Jon Crisp has a strong track record in securing contributors for documentaries on difficult topics and has the knack of persuading people to appear on camera and breaking down barriers. He advises, 'If possible get to know your contributors off camera before filming starts, but let them get used to talking to you with a camera in your hand and the trust will come. A good tip is to give them something about yourself, your own personal life or the reason why you are making this film about them. Be as open as possible with them, that way you have a good natural two-way relationship from which you will get really good dialogue and invaluable sync.'

'I call it the Oprah school of interviewing,' says US based producer Richard Drew. 'Reveal a few things about yourself and people often open up about themselves.' Be willing to open up and share confidences – but not too much; remember, they're the subject not you!

Like celebrities, members of the public also need reassurance and support. By giving positive feedback you can boost their confidence. 'Most people won't ask but *everyone* is dying to know how they come across on camera,' says Jon Crisp. 'So after the first bit of filming help them relax into it by telling them they're coming across really well and giving you some really interesting footage.'

Among the sensitive subjects he has covered, Jon has persuaded people with eating disorders to take part in a documentary. He has a number of suggestions for keeping difficult contributors on side. 'Find a reason for them to want to do it,' he says. 'Tell them that their opinion or view on the subject is invaluable. If they are pulling out of filming, quite often they just need to throw a tantrum and get something off their chest. It is essential to film it as that will be some of the best raw sync (soundbites/speech) you'll get, so tell them it's important you film it as it is an important part of their journey and it needs to be covered.'

Richard Drew has worked primarily on entertainment and reality shows, but similar rules apply. He says, 'Be direct. Ask them what their concerns and questions are and then address these. People respond well when you're honest with them; don't lie and tell them you're going to do something if you're not.'

Interviewing

Interviewing is an art. Get it wrong and it can be boring. The researcher's interviews, whether in finding stories, facts or talking to experts, will crucially affect the content of the programme. It's important to get first hand comment, personal opinions, experiences and facts that only the interviewee can give you. Too many questions and you're like an interrogator, too few and you don't get the full picture. Your interviewee must have memorable and interesting stories to tell that can contribute to or make an entire programme.

You need to be polite but tenacious. If you need to get some information be persistent, don't give up until you've got it. Flatter but try not to be the person's friend. Take your cues from the person you're interviewing – watch their body language and how they react. Know when to ask questions, when to listen and when to be silent. Often this can yield results as they want to fill the silence and might offer you information!

The key to doing a good interview is preparation. Know your subject in advance. Find out as much as you can before you even talk to someone. Research your area and subject well so you know what you are talking about. Always make sure you are well briefed.

'Always arrive fully prepared,' says Rachel Roberts. 'This means doing your homework. Always do plenty of research. Take and make notes so you can refer to them before you start and during the interview. Dropping in an obscure but meaningful piece of information about a celebrity can often do the trick. It shows you've done the research and makes them feel appreciated.'

Decide what you want to get out of the interview and lead the person in that direction. Know what you are going to ask and when. Prepare a list of questions before you start the interview. But do be prepared to go off the subject to follow interesting avenues of conversation and areas you might not be too familiar with.

If you are interviewing a celebrity before they are due to come on your show, make sure you're aware of the content of previous interviews so you know what has been covered to death and what people might have already seen, and what is interesting and what's not.

Simon Warrington says, 'If the questions are about a guest due to appear on the show, I obtain old interviews from various other shows, or from the previous show if they're re-appearing, and write down their questions and answers. The last thing a guest wants to be asked is the same old questions, nor does the presenter want to go over old ground.' Looking at old interviews will also give you leads into other areas of interest that might lead to an exclusive.

If you are doing a pre-show interview on the phone think about your phone manner and technique. Smile a lot. Don't be abrupt or rude, try and keep warmth and familiarity in your voice and your manner so the person you're talking to can relax. Be friendly and open, share stories and hope that your subject will too. Engage your subject by being charming.

'Whether I'm interviewing "ordinary" folk or celebrities, a bit of charm goes a long way,' says Rachel Roberts. 'But don't do it in a disingenuous way or your interviewee will not trust you. That's the most important thing you have to achieve. If your interviewee is mistrustful, it will be a stilted experience.' Be polite and mindful of the fact that your agenda is not theirs.

In interviews:
• Your first question should be 'Is this a good time?'

- Check you're speaking to the right person and check their name and title
- Never ask closed questions – where the answer might just be no or yes. Make sure you ask open questions where someone is forced to give you a detailed answer
- Remember to ask obvious questions the who, what, when, why, where and how. Always check your facts – the most basic things about the person, who they are and where they're from

Rachel Roberts says, 'I learnt the five Ws – who, what, why, when, where. I *always* check name spellings and ages. You also have to maintain your curiosity about the world around you – you've got to ask questions to unearth stories.'

Write a list of questions for your interview and structure so it has an order and is in sections or blocks of topics so you know what you want to cover and in what order – though be prepared for things to change in the interview and to go with the flow.

Always lead with the easier, softer questions. Save the more difficult questions until last when you've warmed up your subject. If you don't get a satisfactory or an explanatory answer, park the question and ask it in another way later. 'Don't be afraid to ask the same question more than once, only in slightly different ways,' says Lucy Reese. 'It's always useful to have a choice when you are in the edit.'

There are subtle ways of asking the same question differently that will result in different answers. Also, as you warm up your subject, they might relax and feel comfortable about sharing information with you.

If you can't successfully interview someone on the phone and get what you need, try to meet them face to face. Making personal contact will break down barriers and inspire trust. Arrange a face to face meeting so they have a chance to get to know you; take someone more senior with you so they feel you are making an effort with them and that you have back up and support to persuade them you're genuine, trustworthy and from a reputable place.

'It's all about trust,' says Rachel Roberts. 'People generally warm to me as a person, so it's not that hard to get them to open up. Use your full armoury of people skills to get what you need. Make it feel like a chat and not a formal interview.'

Don't expect your interviewee to perform unless you've put in the groundwork and know about them and their situation, showing you've done research will go a long way.

Interviewing on camera

If you have found a contributor and nurtured them, often you will be asked to interview them for the programme as you have the relationship with them. As with celebrity interviews you usually won't be seen nor your voice heard, just the person's answers edited with pictures and a voice over.

If you are interviewing someone off camera you need to bear in mind different rules and be aware of different things. During the recording you need to make sure you get coherent, useable, relevant sound bites that can be used to voice, illustrate and narrate the story. You need to make sure your voice doesn't cut off the interviewee's answers or interrupt them mid-way through a sentence. Wait to ask questions so you don't cover what they are saying and allow them to finish what they want to say. Don't be afraid of silence – give them the space to carry on without interrupting them with your next question. Allowing them to lead the interview might reveal fruitful and undiscovered areas as they give you information.

It's important that you make sure your interviewee is relaxed, briefed and knows what is expected of them. You must make it clear that you are not going to be on camera so they know they must give complete answers (as your questions will be edited out).

Asking an open question with the prefix 'tell me' is always useful, for example, 'Tell me what happened when you arrived at your office that day.' By using 'tell me' you are inviting your interviewee to give you a full answer rather than a brief one.

You should write a pre-edit script before you do your interview so you have a plan for your material and know what you require in the edit for your show. That way you'll focus on getting the answers you need.

'Think about your edit,' counsels Richard Drew. 'It's the old golden rule of putting the question in the answer. You have to imagine how all the sound-bites are going to be cut and what you need. That comes from experience and being prepared. That's when a shooting script can save you. Celebrities are well honed and will give you nice, clean answers but members of the public will need some coaching. But if you go in with a clear plan of what you need from the interview you'll be fine.'

Writing a pre-edit script will help you draw up your list of questions. These should be focused on different areas and will help structure and order the interview so you avoid jumping around and can follow through on your topics.

Always write a list of questions and take them into the interview with you, even if you feel you don't need to. They are useful to refer to and to keep yourself and your interviewee on track.

'Over the years, I've become more adept at compiling a mental list,' says Rachel Roberts. 'Although I will write down a few pointers to glance at on my way. A pretend list of questions can come in handy if you have a particularly verbose subject! Looking at your sheet of paper is quite authoritative and will put you back in the driving seat.'

Never rigidly stick to your questions. Make sure you ask them all but allow your interviewee to say what they want to say. They will have their own agenda or message to get across. Let them do this, listen with interest and ask them about their subject. Once they've got it off their chest they'll feel they are in control and you can steer the interview to where you want it to go.

Try to make your subject feel relaxed by being relaxed and comfortable yourself – even if you are probably under pressure of time and might feel nervous. Make sure your body language is open and warm so they open up to you. Be confident and radiate charm and calm (even if you don't feel it!). They'll gravitate towards you.

I used similar tactics when I did interviews at press junkets with movie actors promoting their films. I was one of many other shows and channels interviewing the celebrity on the same day. They'd fly in for two days and rattle off any number of interviews with newspapers, radio journalists, TV shows and also press outlets abroad. They remain seated all day in the same spot while I took my turn in an allotted 10 minute spot (which normally has been reduced to about eight minutes – sometimes less due to over-running). You are not supposed to overrun and have some desperate PR person waving off camera, telling you that you have three minutes left, one minute left and waving their hands around telling you to wrap up the interview.

It's incredibly distracting and stressful and does little to create a relaxed atmosphere, but when I did them I would try to be as relaxed, charming and confident as I could. My aim would be to make the actor feel at ease and to have a laugh with them so I would get good material and responses from them. I would compliment something I had seen in the film and rib them about an aspect of it or make a funny observation. I told Michael Caine I'd heard that he had tried to keep the hideous, pimp style clothes he wore in the movie *Little Voice*, which luckily he thought was funny.

'People almost always respond to a smile and someone who is nice,' says Richard Drew who has done hundreds of junket interviews. 'Junkets can be relentless for celebrities, nonstop interviews all day long, but if you bound in friendly and energised you get much better results. I've had celebrities compliment me and say how much they've enjoyed chatting at the end of a very long day.'

Rachel Roberts always tries to put her subject at ease. 'I'm more confident and relaxed, which generally rubs off on my subject. If something doesn't go quite according to plan, I move on and go down another route. I'm also bolder, asking those tricky questions without a second thought.'

Don't be afraid to ask difficult questions. Be genuinely interested in your subject – try to ask the things that viewers would ask and want to know. Be curious and learn new things.

Lucy Reese suggests these golden rules:
- Ask the nice questions first, eg, 'Tell us about your latest project' rather than 'Tell us about your divorce'.
- Don't be afraid to grovel – the person you are interviewing is doing you a favour. But don't flatter too much. Interviews should be challenging.
- Keep the interview as short as you can – 1 tape ideal, 2 ok, 3 getting a bit dodgy. Even people who like the sound of their own voice will get tired after more than hour of continuous questioning.
- Don't clip the end of answers however tempting it is to move an interviewee on.
- If you are unhappy with an answer, because either you or the contributor has made a mistake, say, 'We need to do that again for sound'.

Don't be afraid to stop recording and do some of the interview again, if you need to – if the person is tense and their answers are clumsy and don't make sense, for example. Blame a glitch with the recording – the picture or sound – to make sure you get what you need to make a good show or interview work. You may find that they might be less tense the second time around and you'll get a better answer.

If the interview is going badly there's always a risk that you might not be able to turn it around but by taking a break, checking that they are happy, asking them if they need anything, or allowing someone to vent their anger and displeasure are good ways of getting it back on track.

I've often found that if you don't argue or contradict someone when they are raging, after they've got it off their chest and have calmed down most people will feel guilty for being so mean and will apologise and try to make amends. It's not nice being at the receiving end and it can be difficult to bite your tongue – especially if it's unjust – but being silent and taking it usually pays off. Just allow them to finish their rant and to slowly come down and they will.

Lucy Reese has these suggestions for dealing with a difficult interviews:

- Flattery
- Ask if they want anything – food, drink, extra make up etc
- Offer to show them the shot and tell them how wonderful they look
- Make sure they sign the release form – if they make a fuss about specific clauses, tell them to cross them out and initial them. A limited release is better than nothing
- Tell them they don't have to answer all the questions you're asking

You could always give them a fact from another person to comment on. 'You have to be a little bit crafty and warm them up,' says Rachel Roberts. 'Feed them a quote for them to agree or disagree with. That way, even if they say yes or no, you have something you can use.

With celebrities the PR will often warn you in advance to avoid certain topics and not to ask questions in certain areas, but there are ways around this.

'One tip I used to find helpful when doing junkets was if I had to ask about a touchy subject, a celebrity's romance, bad reviews for a movie, etc, I'd ask the question at an angle,' says Richard Drew. 'I'd say, "Some of the press have been a little sniffy about your movie, how do you respond to that?" Often the celebrity would end up talking about the subject even though they kind of didn't want to – and you've not been antagonistic when asking the question. You've placed the blame elsewhere. It worked for me!'

- Make sure any difficult questions follow some soft soap
- Send flowers to the agent to say thank you afterwards and they will be putty in your hands forever

A thank you letter, follow up email to their agent or a call to them can all make a difference and hopefully sow the seeds for a relationship that you can call upon in the future.

How to write a brief

Once you've done your cuttings research and phone interview (if it is not on camera but in preparation for your producer or presenter) you'll need to write a brief. This should be no more than two pages long and should be written in a clear, simple way with all the salient and crucial facts included with no spelling mistakes.

'A brief should be precise, to the point, not full of waffle,' says Richard Drew. 'It should be well laid out, no typos or spelling mistakes. I don't want to have to spend my time correcting researchers' mistakes which they should have spotted themselves.'

A brief should be exactly that: precise and concise. 'A brief involves giving a summary of the subject matter and allows the presenter to know exactly what is going on the minute he or she reads it,' says Lucy Reese. 'It should be informative and concise; no one wants an essay.'

It's a supporting document for the person doing the interview and they'll need to be able to digest and take in all the relevant information clearly. Researcher David Minchin prides himself on presenting facts in a digestible manner. 'Get the facts then write them out however your director or whoever may need them, he says.'

Make sure that you check your facts and don't make silly mistakes. Lucy adds, 'Make sure all the information is 100 per cent relevant to the programme item. A presenter does not have the time to absorb unnecessary facts. They might not even have time to give the brief more than a quick, scanning read so make it as clear as possible.'

Write it in a simple and clear way. Don't make it flowery or too colloquial – just straight, to the point and grammatically correct. The most important thing is to assume your reader knows nothing about what you're talking about – it's then up to you to communicate the information so they understand what you're talking about immediately. 'Try to write as you speak,' says Lucy Reese. 'The best written communication is direct, punchy and personal and doesn't show off about how clever it is.'

Again, who, where, when, what and why are good to bear in mind when writing a brief, or any written communication. Always use the active tense where possible and avoid long-winded sentences full of semi-colons and long words. Write in short direct statements, leading with the most important headline grabbing information first to hook your reader, this also applies when you are structuring a programme narrative and is a good discipline to learn now.

Make sure the information is relevant to your viewers. 'Who is the audience? A GMTV viewer or presenter would want a different take on the story to a Newsnight audience,' adds Reese. Similarly remember the age and outlook of the show's target audience. Different channels and shows have different demographics. You should be familiar with the needs, aspirations and desires of your audience.

Setting up shoots

A researcher is responsible for setting up shoots, finding locations for interviews and making sure they are cleared, that guests and locations sign release forms permitting their use and agreeing to the company's terms in writing.

A researcher will usually carry the location and contributor release forms and the new tape stock. They are sometimes responsible for labelling tapes and ensuring that the raw footage of the shot programme (known as rushes) are safely taken back to the production office at the end of the shoot and placed in the editing suite.

Researchers have the support of the production management team in putting the call sheet together: the production secretary, the production co-ordinator and production manager help with booking the crew, equipment and making the travel arrangements. However, with so many people involved, the researcher needs to be meticulous in checking details as they are ultimately responsible for everything.

On a studio show the call sheet is a bible which includes the show's script. On location it can range from a page to three to a booklet.

Shoot days are expensive as companies often have to hire kit, camera and soundmen, lighting, transport and locations and/or a studio for a day, so they'll want to make sure that every available minute is used to justify this expenditure. A shoot day will often involve different locations, multiple set ups and different arrangements. Usually a researcher will have been on a recce before the shoot takes place to work out what is going to be filmed and where and with whom.

The call sheet
This is an essential document containing all the information necessary for everyone involved in the shoot.

It contains:
- Contact details for all the production team members working on location, and the production management team in the office
- Contributor and location contact details – addresses and telephone numbers
- Call times for all the crew and rendezvous (RV) points
- All the crew's travel details for the day's shooting, when they are being collected and expected at the first location
- It will include the amount of time required to set up and the amount of time estimated for the shoot to take place (known as the record – RX) and the final wrap time of the shoot (when the day is officially over). It is normally based on a 12-hour shooting day
- Scheduled breaks
- Details of where the crew can park and can eat
- The company's insurance certificate
- The risk assessment – a document which outlines the potential health and safety hazards faced in shooting the production and the way in which the production team will minimise these risks and deal with them if they arise. The BBC offers specialist training in health and safety which is used by most production companies. The risk assessment is usually written by the series producer with the production manager.

- Maps of and directions including details of the nearest hospital and police stations and hotels
- The pre-shoot script – usually written by the producer/director or series producer or producer, which outlines the content expected of the shoot and how it will fit in the programme

A sample call sheet could look like this:

<table>
<tr><td colspan="3" align="center">**Company name and contact details**
Show name
CALL SHEET</td></tr>
<tr><td>NAME:</td><td>TITLE: Executive Producer</td><td>MOBILE NO:</td></tr>
<tr><td>NAME:</td><td>TITLE: Series Producer</td><td>MOBILE NO:</td></tr>
<tr><td>NAME:</td><td>TITLE: Presenter</td><td>MOBILE NO:</td></tr>
<tr><td>NAME:</td><td>TITLE: Producer/Director</td><td>MOBILE NO:</td></tr>
<tr><td>NAME:</td><td>TITLE: Assistant Producer</td><td>MOBILE NO:</td></tr>
<tr><td>NAME:</td><td>TITLE: Researcher</td><td>MOBILE NO:</td></tr>
<tr><td>NAME:</td><td>TITLE: Runner</td><td>MOBILE NO:</td></tr>
</table>

PRODUCTION CONTACTS

NAME	TITLE	CONTACT DETAILS
	Production Executive	Home number Office direct line Mobile number
	Production Manager	
	Production Co Coordinator	
	Production Secretary	
	Runner	

Technical requirements
A list of camera equipment required
The camera kit and cameras being used
Lenses
Lights
Sound equipment
Number of tapes and quality, for example 6 × 30 beta sp tapes

Other requirements
Any special lenses or equipment, for example, a wide angle lens for effects, dolly (a track that the camera moves along), filters to affect light and image quality

INTERVIEWEE CONTACTS

NAME	CONTACT DETAILS
	Including their home numbers, mobiles and full address

LOCATION DETAILS

NAME	CONTACT DETAILS
Location 1:	Full address and telephone numbers plus name of the contact the venue is booked through
Location 2:	
Location 3:	

DETAILED ITINERARY

DATE	Monday 15th September
05:00	Researcher briefs contributor 1 and recces location
06:00	PD and AP, researcher, contributor and runner RV at location 1

06:00	PD and AP set up and light: sync cameras and prepare for interview 1
06:15	PD and researcher discuss
06:30	RX: PD shoots the interview with contributor 1 Researcher shoots camera 2
06:30	AP shoots additional material and GVs outside
08:00	De rig (pack up and prepare to leave for location 2)
8:30	PD and researcher depart for and travel to location 2
8:30	AP goes ahead in separate vehicle
08:45	PD shoots interview with contributor 1 en route to location 2 with researcher AP shoots exterior shots outside
09:00	AP arrives at location 2 and sets up for arrival
09:00	Team arrive at location 2
09:00	AP shoot arrival of contributor 1 at location 2 and films sequence
09:30	PD and researcher set up interior for interview 2 with contributor 1

The call sheet will carry on in this way until the shooting day wraps.

MAP AND DIRECTIONS

LOCAL HOSPITALS AND EMERGENCY SERVICES

USEFUL NUMBERS

For example, local taxi numbers
local restaurants, etc

Working as a shooting researcher

It's becoming increasingly common for researchers to shoot material for transmitted programmes and shoot tapes promoting new programme ideas in development. Often researchers will shoot on location providing material in their own right or as a supporting second camera. They may be responsible for lighting, sound and framing their subject and also doing interviews and shooting general views and sequences often on digital handheld cameras like the Sony Z1, the PD 150 or a DSR500.

'The earlier people pick up a camera the better,' says Julia Waring. 'It's good to get as much practice as you can. Learn how a camera works; it's a useful skill to have, especially as more and more production companies are hiring self shooting producer/directors and assistant producers who produce and shoot all their own material for a programme.

Being able to shoot is becoming essential. Increasingly programmes are entirely shot on digital handheld cameras by the teams who make them – producer/directors, APs and researchers.

After just two years in television Jon Crisp was a shooting researcher on a five show. Entirely self-taught he could competently self-shoot on Z1, PD150 and PD170 and had a good deal of location experience. 'I'm a bit of a gadget geek which helped getting to know how it worked,' he says. 'You need to be able to operate the camera and have a good idea about angles and eye lines. Naturally, with the shooting responsibility comes the need to be able to interview and get good sync from contributors. I think it's essential that you can concentrate one hundred per cent on content as you don't want to be learning to operate on the job. Get to know the camera first.'

During an intensive seven week shoot he was one of two main camera operators for the 5×60 series and was given the responsibility of filming and directing a number of his own shoots. 'I learnt how to shoot because I was interested in cameras and filming,' he says. 'I took a camera home and played with it and practised on friends. I learnt to shoot simple sequences and figure out what shots you need to cover it in the edit, including big wide shots, listening shots and establishers.'

Being able to shoot well lead to his first credit as an AP – still just two years into his TV career. 'It is very good stepping stone to get from researcher to AP. So long as your rushes look good and the content is usable you won't spend much longer researching before you get an AP position' he says.

Jon is now in much demand because of his talents as a self shooter and has worked at RDF, Windfall Films and Betty. 'He has managed to climb up the TV ladder with relatively little experience because he is a fantastic shooter and a good researcher/AP,' says Emily Shanklin, head of talent at Betty. 'We hired him as I thought he had an easy going charm in the interview that would go down well with contributors. Trying to cast difficult contributors is our bread and butter so I thought he had good skills. Not only did he find the people prepared to talk about their experiences but he shot a fantastic taster tape for the development of the show which was beautifully framed and produced. He'd put a lot of thought into how he would bring the interviews to life. The result was that we won the commission. We then hired him as an AP on a documentary for C4, despite that fact he didn't have a lot of documentary experience because we were sufficiently impressed with him on the development.'

David Minchin has also taught himself how to shoot. 'I got lucky and got my break. I've always been interested in video cameras and when I was working on *WAGs Boutique*, because of the nature of the show, there were hundreds of things going on at one time. There were lots of Z1 cameras lying around and quite often we were short staffed or things that happened outside the shops where we were filming and they would often need someone to film the WAGs going to get lunch or getting stock. Because there weren't many hands they asked if anyone wanted a go and I said "Me! Me! Me!" '

Although he was a novice he had the confidence to keep trying and learning as quickly as he could. 'First of all my shooting wasn't great but the more you do it the better you get. I started doing more and more and eventually someone said my stuff was good, so I kept asking to keep shooting when I wasn't busy and they said yes. I shot with a DSR on *Wish You Were Here?* I was lucky, I had a lovely director who liked me asking questions.'

David has learned everything about shooting on the job. He says, 'Always ask questions, at the right time and intelligent questions; there is nothing wrong with asking someone why they are doing something or why are they asking them to do something again or taking it from another angle. The problem is that no one really trains you in TV. I've only ever had health and safety training and the rest of it I have picked up from asking questions and seeing if I can take a camera home and have a little play.'

Claire Richards says, 'In almost every job I've had as a researcher I've had to shoot.' It's a useful skill which has provided opportunities and she believes it has ensured that she's worked continuously and given her an advantage in the job marketplace. She explains, 'You do need to be able to shoot as a researcher and AP. It helps you and gives you a bit of an edge. If you have an amazing interview that you need to film and you can't get a camera, it's great to be able to do it yourself. On my last production we set everything up and shot it ourselves, and I shot extra shots for the edit.'

But it is not enough to be able to competently use a camera, the skill is in the rapport and the relationship you have with the contributor and the performance you get out of them, and constructing a narrative and telling a good story.

'We decided to devise our own course with the dual aims of ensuring, first, that there was a balance between the technical and editorial, and second that the tutors were all leading DV self-shooters. The courses were also designed to be incredibly hands-on with lots of practical exercises, so that, as far as possible, mistakes are made in the classroom rather than out in the real world. The formula has been a huge success and is one that we have carried into other areas as we've expanded our training offering.'

Libel law

Every researcher working in TV should have a basic knowledge of law, although most people don't have any formal training. You are simply expected to know, though great problems are caused by legal difficulties that programme makers are unaware of.

When I trained in journalism I was taught the basics and used a text book called *McNae's Essential Law for Journalists*. This is essential reading for journalists in print, TV and radio. It's accredited by the National Council for the Training of Journalists and is worth investing in.

Ultimately, the series producer is responsible for ensuring the programme meets standards in compliance and broadcasting, and they can call on advice at any time as all production companies and broadcasters have lawyers who can offer guidelines. But it is good to have a working knowledge of libel law and good practice so you can raise any issues that might come up. I know of documentaries which have been unable to be transmitted because simple and basic procedures were just not followed – like getting a contributor to sign a release form.

There are strict codes of practice on what products you can show or 'place' in a programme, in what context and how you use, clear and credit archive clips, and on how you represent and show people. It is good to know what these codes and rules are.

Defamation

If a broadcaster publishes something which is defamatory they commit libel. It is possible to libel individuals, groups or organisations – and the risk exists whether the defamatory statement is scripted or spoken off the cuff. The broadcaster is liable no matter who speaks the words in a programme or whether it is made in house or by an independent production company.

There are a number of tests applied by the courts to determine if something is defamatory and there are three principal defences. They are complicated and apply to a wide range of programmes.

Any programme maker can consult the company's or broadcaster's lawyer at any stage of the production – the earlier the better. They will be able to give advice to help them minimise the risk of libel and to avoid breaking the law. In the light of the TV scandals that hit the industry in 2007 transparency and viewer trust are key issues so knowledge of editorial policy is very important. Ofcom offers advice on its website and information for broadcasters including the Broadcasting Code.

PROFILE

Tim Hincks, chief executive officer, Endemol UK

Tim Hincks is responsible for Endemol UK's content and its production brands – Initial, Cheetah Television, Zeppotron, and Brighter Pictures. Its broad range of programming covering reality (*Fame Academy, Big Brother*), factual (*Sex Education, Supersize Vs Superskinny, Restoration*), comedy (*8 Out Of 10 Cats, Law Of The Playground, Would I Lie To You?*), entertainment (*Deal Or No Deal, The One And Only, Golden Balls, 1 Vs 100*).

Tim began his television career in 1990 when he divided his time between producing BBC 2's *Food and Drink* programme for Bazal (now Cheetah Television) and working on current affairs programmes such as *Newsnight* and *BBC Westminster*.

He was appointed creative director of Endemol UK in 2002 and stepped up to become chief creative officer in January 2005.

In 2008 Tim was appointed chief executive officer of Endemol UK. He sits on the Endemol International Board and is the executive chair of the Media Guardian Edinburgh International Television Festival.

What does your job involve on a day-to-day basis?
Running the company, selling programmes, creating programmes, surrounding myself with people much cleverer and talented than I am. And taking all the credit, obviously.

How did you get your first job in TV?
I did a very simple and obvious thing. I got a copy of the *Radio Times* and I looked at the names of people who made television programmes and I wrote to about ninety of them saying that I was interested in working in television. Two of them got back to me. One was the editor of Panorama and one was the executive producer of *Food and Drink*. I ended up getting a job on the show. It all seems very random. The notion of training schemes didn't enter my head. I didn't know about them and I didn't bother looking. It was a lot more of a free market than it had been for generations before me.

How did you get your first break?
The first thing I did was research a book on food which the executive producer of *Food and Drink*, Peter Bazalgette, was writing. I did a three-month attachment writing the book. The deal was that if it went well I would be taken on for the television programme. I didn't know anyone in television. I was 21, I'd recently left university and I didn't really have much money. I remember thinking at the time that it was a risk. I felt that I was giving up my temping job for a three-month contract which could lead me nowhere but I thought it was worth having a go. I felt that there was probably quite a good chance of getting a job on the show. It went well so I got a job as a researcher on *Food and Drink*.

What qualities do you think helped you get on?
I felt at the time pretty lucky. What was interesting about that
particular job was that it was actually a factual show and it was in
between genres. It was sort of food journalism, it was looking at
food scares and what's in our food. I'd worked as a researcher for
the National Consumer Council as a temp, I could say that I was
able to do research. Most people start as a runner so it was pretty
unusual, I think, and I felt it was a really lucky break.

So how did you progress your career?
I made my way up the ladder on *Food and Drink*. I became the show's
producer on the second series. But to be honest I'm not quite sure,
I was always quite interested in politics [he did a politics degree] so
I was thinking quite seriously about it. I was applying for other jobs
all the time during the second half of the first series of *Food and
Drink*. I flew loads of CVs off and of course the brilliant thing that
happens to you is that once you're working on a TV show, the letters
you send with your CV suddenly get read. With your first job you are
able to get a foot in the door. You are treated in a very different way
from the mass of people trying to get in. I actually got offered a job
on a political current affairs programme which I was very keen to do
at the same time as being offered the job as a producer on *Food and
Drink*. It was a toss up between doing something interesting or doing
something ambitious to boost my ego. No contest obviously.

So I took the huge step of being a producer on *Food and Drink*
and I wanted to take a job higher up the tree rather than regret
not doing it. I ended up as the series producer and the executive
producer on *Food and Drink* over the course for four years, but in
between those contracts I would go off and do a job somewhere
else. I went around the world and spent loads of money travelling
for a few months and then during another gap I worked on
Newsnight.

The fact that it's a freelance industry means you can play it a bit
and not get too worried about slotting in to a certain path but I

had a certain faith that *Food and Drink* had some sort of security because it was being commissioned each year. I then climbed up the ranks and grabbed the opportunity. But equally I was having fun freelancing for six months when it was off air. Once you have got work as a researcher or producer you become more valuable.

Did you have a mentor?
Peter Bazalgette gave me my big break and was a huge support. I learnt a lot. God knows what became of him.

What advice would you give someone wanting to have a TV career?
Always smile, and don't be an idiot. You'll stand out a mile.

Did you have a gameplan?
No, I'm not sure people have game plans. I think it's more about having a sense of what feels right and what doesn't. It's not about thinking that you'd like to be a producer, executive producer by the time you're 28 or whatever. You just know when it's right and if you're lucky enough most of the time the opportunities feel like breaks. I feel *Food and Drink* was a brilliant opportunity. The game plan was to try to do some interesting things and it was an alternative to doing a real job which is what telly is.

What qualities do you have that have helped you in your career?
I find the process of creating formats, entertainment formats, is something that I enjoy and I can do. And I can work with groups of creatives and develop stuff into formatted programmes. There are very few people who can create ideas from scratch and actually very few people enjoy it because it's so frustrating.

What's the best thing about working in television?
In a very pretentious way I like the feeling of being part of the national consciousness. What we do does have a cultural value and I love arguing that, it's great fun. Clearly, there is some kind of

underground ego that I'm trying to control because what I do goes out in the public arena and is hugely talked about and sometimes rates highly. But quite often our shows collapse and are disastrous. You take it on the chin because the viewers are always right, no matter what critics will tell you.

What's been your best TV moment so far?
Bobby Davro doing a belly flop live on *The Games* on Channel 4. No contest.

And the worst?
It's genuinely been a lot of fun so far. We're all very lucky. Teaching or working in a hospital, now that would be tricky.

Chapter Five
Plotting a career path

How do I plot a career path in TV?

There is a career path in TV production for creative editorial talent. As we have seen, new entrants normally start as a runner or junior researcher, progressing to researcher through to assistant producer, then producer and/or producer/ director, to series producer and series editor and finally to executive producer, the top position on the TV production ladder. (See p. 205)

'As a researcher there is a lot you have to learn about how a programme is made. You learn about responsibility, trust, what's important and why,' says Moray Coulter. 'As an AP you learn how to crossover between a senior researcher and a junior producer director. Once you have worked as an assistant producer you are ready to become a better producer/producer director or director.'

Although the runner is the lowest rung of the ladder and the executive producer is the highest, there are no hard and fast rules about getting from the bottom to the top. Career progression can be fast, especially when you are freelance – you are in a position to negotiate your rate and position and if you have worked on a hit programme to earn promotion quickly.

'People can be very impatient to move up the ladder as quickly as they can,' says Coulter. 'You can hear about a researcher who has gone up to be a PD in a year. But it's not good for the industry for people to move up that fast because you have to learn so much on the job for each particular responsibility.'

Promotion is faster than it ever used to be and younger people are less keen to hang around in what are deemed to be junior roles, but it is essential you spend time picking up new skills and learning from those above you.

Career Trajectory

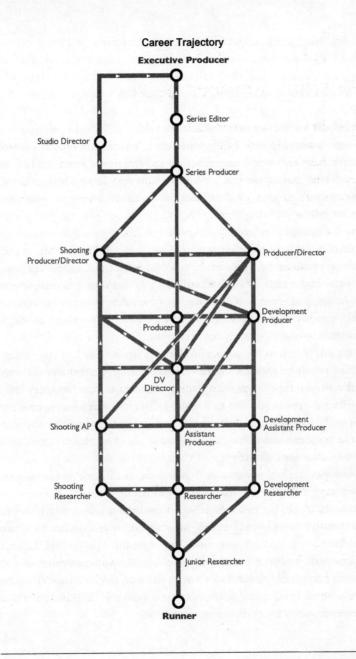

Executive Producer

Series Editor

Studio Director

Series Producer

Shooting Producer/Director

Producer/Director

Producer

Development Producer

DV Director

Shooting AP

Assistant Producer

Development Assistant Producer

Shooting Researcher

Researcher

Development Researcher

Junior Researcher

Runner

Nick Holt got his first break and gained his first PD credit on a BAFTA nominated documentary for five called *Guys And Dolls*. He says, 'I think all the skills that helped me make *Guys and Dolls* were learnt whilst I was an AP. So don't underestimate that AP role.'

He believes the best way to sustain a career in TV is to progress at a steady speed. He says, 'Take your time. I think there are a lot of people in a mad rush to be a director and I'm never sure why. The AP role is vital and allows you a great perspective of the whole process. You can learn a great deal without the pressure of being responsible for everything.'

You shouldn't overreach yourself too fast because if something controversial happens on a programme or you get into a difficult situation it may be because you haven't had the training, background or experience to be able to handle it. 'You can't jump a rung of the TV ladder with every job you take; be patient and work hard in every job,' says series producer Anna Blue. 'Don't expect to become a producer or a series producer overnight.'

'It's important to know your level to the absolute best of your ability before moving upwards,' agrees Louise Mason, a producer director and edit producer. 'Spend a decent amount of time at every level before progressing to the next role. I think it's really important not to run before you can walk in television. Many people are eager to get to the next step as quickly as possible, but I wanted to make sure I had a thorough knowledge of the level I was at before moving on. I had good role models and teachers, and took their advice, encouragement and criticism very seriously.'

Like Louise, who switches between production directing and edit producing, most people in TV production alternate between two different job roles and have more than one job title on their CV, for example a junior AP will still be available and consider senior researcher roles. 'Never worry about titles or credits,' says PD Anna Keel. 'A researcher position is better than an AP position if the former involves more directing experience.'

The lines between roles are blurred as chance, circumstance and opportunity play a huge part in accepting roles. An AP might be able to work as as a junior PD, or a producer, edit producer or a researcher depending on when they are looking for work and what is offered at the time.

The majority of people in TV are freelance, there's no holiday, sick pay or overtime. You only get paid when you are employed, so you don't want to have too many unpaid days between contracts, or to take on a lesser role to the one you were on previously. 'I'm freelance and choice isn't always a luxury that can balance out paying the mortgage,' says Louise Mason. 'I think remaining flexible is the key to surviving.'

Freelancers have to take opportunities as they come and so an edit producer might take work as a PD, or an AP might take researcher work if they like the project and they have few alternatives at that time.

It's a good idea to think about what your long-term goals are and how to achieve them, and try to balance the need to eat with the credits on your CV. 'It's important to use your critical faculties before accepting a job and, if finances allow, not just take the first thing that comes along,' says Matt Born. 'Have a goal of where you want your career to go and what kind of programmes you want to work on and keep that in mind at all times. Before taking a job you should ask yourself whether it's a step in the right direction. What channel is it on? What slot, i.e. profile, is it going to have? And what budget? Is the company making it well thought of? Do they make the kind of programmes you want to be working on? A job interview shouldn't just be about the production company asking whether you're right for their project; you should be asking them why their project is right for you.'

There are many deviations from a traditional A-to-B route of career progression in TV. There are no hard and fast rules because of the role contacts, networking and mentors play. It depends on the relationship you have with the person who's doing the hiring. If someone likes you and rates you and is in a position to promote you, you'll get bigger breaks and opportunities.

Most entrants get in because they know someone and once you're in it's about making strategic alliances, friendships, networking and working with the right people and doing a good job. There are executive producers who were producer/directors who by-passed the series producer role because they were promoted by someone who knew and trusted them and had faith in their abilities because they had already worked with them or because they were carried on the wave of a hit.

'In any genre, there's a career path, say researcher to AP, etc,' says AP Harjeet Chhokar. 'Within that, you can choose just one programme and, usually very quickly, progress through that. For example, there are people I know who have only ever worked on *Big Brother.*'

These people have worked on a long running series for their entire career – getting promoted with each series – and end up running the show. But it's not a good idea to just stick to one show and have nothing else on your CV. You need to make different contacts and work in different companies to increase the size of your network. If you only work on one programme you'll only learn about one particular style of programme making and it could hinder your chances of getting work outside that indie – and it's never good to be reliant on one employer. Learn as much as you can and once you've done so, head for pastures new and find and take new opportunities.

'As a freelancer, it is important to make as many contacts as possible and to work at different companies. It's a very bad idea to work on just one programme,' says Harjeet Chhoker.

'I'm slightly suspicious of people who stick with the same show for a long time,' says Richard Hopkins. 'Like people who have gone from being a researcher to executive on *Big Brother.* If I got a CV through the door like that I would slightly worry that they were reluctant to come out of an institution and would find it difficult to work in a new environment.'

At the beginning of your career it can be useful to stay at one company. Especially if they offer you the chance to learn new skills by working on different productions. You'll learn a huge amount on the job and it will offer you stability and a regular income. However, TV is a freelance world and it's good to spread your wings and try other places. It's good to get out and network and become known and establish your reputation.

TV is a meritocratic industry but those who prosper do so not because they are the best or the most experienced but because people in positions of responsibility believe they are smart because they've worked with them or because of their track record and recommendations of people they know and respect.

But a successful career has as much to do with luck, chance and getting in with the right people at the right time on the right shows as much as it has to do with hard work and talent. Get on a hit show with a successful production company and deliver and you'll be in demand. Once you've worked on a successful show your phone will start ringing and job offers will roll in.

'Having a mixture of hit shows on your CV is a clear indicator that you've either got a nose for a good show or you've been a part of making it,' says Richard Hopkins. If you choose your roles strategically you'll go from hit to hit and rise meteorically. Katie Rawcliffe was an executive producer by the time she was 29. 'I was ambitious,' she says. 'I really wanted to do well and I was always excited about doing the next thing but I never actually applied for a promotion. I applied for jobs. It's not a case of seeking. It's usually people who I've worked with who must have thought that I can do the next thing which is great. I've never actually said, "Me! Me! Me! I want to do that!"

'It's about how good you are. I've worked hard, I've learnt a lot and I think quickly on my feet. People recognise that so some of it's luck but some of it is reputation.'

TV is an industry where people get on because of who they know and what they've done. If you work on a hit and get a reputation for working hard and delivering creatively, everyone will want you.

To be a good producer/director who's always in demand you have to 'constantly be doing a good job, delivering what's required and then exceeding expectations,' says Matt Born. 'You're only as good as your last credit or, if you're lucky, three credits, so it's important to have a nose for projects that are likely to be successful. Work on the first series of *The Apprentice* or *The Baby Borrowers* and you'll be able to ride that wave for years to come. Of course, finding yourself on a hit show requires some luck.'

Jon Crisp has been promoted quickly and is in demand because he is a skilled shooter whose talents have lead to shows being commissioned. 'I didn't really plot a path,' he says 'I just tried to keep working on projects I believed in and enjoyed, tried hard and the opportunities came along with it.'

He's been in continuous work largely through hearing about new shows 'through the grapevine, previous contacts and heads of talent.' He's succeeded 'by showing willing, that no job is too much trouble and anything I do, I do to the best of my ability. When the opportunity comes up to take on more responsibility jump at it. Most importantly you want people to enjoy working with you.'

If you're associated with a hit show, even if you are a researcher or a runner, you'll get interest from other companies desperate to emulate its success and to hear about your experience. The credits you have on your CV are enormously important if you want to rise quickly.

But not everyone can be a TV trail blazer, there are hundreds of people working in the industry quietly making successful programmes who've manage to sustain a long career as jobbing freelancers, moving from show to show within a genre. Though in these lean times, as more and more people fight for the same jobs, it can be incredibly competitive.

Watch how people succeed in the roles above you and listen to their advice. TV is fast paced and competitive and you need to learn as much as you can to stay ahead of trends, technology and styles of programmes. There is no better place to learn than at the cutting edge – on a programme.

'All my real training in TV came on the job,' says Conrad Green. He says that he learned the TV trade by 'watching, listening and asking people who were much better than me how they did their jobs.'

Having a boss who'll inspire, help and teach you is enormously important. Working with respected and talented practitioners is invaluable experience. They can show you the ropes and share their expertise and, of course, give you work. As successful producers rise you can hang onto their coat tails and rise with them. They might also provide you with news and leads of work elsewhere and recommend you to their friends if they think you're good enough.

There are generally two routes to the creative editorial side of the production process – at researcher level you can either decide to focus on directing or producing. You can choose to be responsible for overseeing the content of the programme rising to a series producer or to direct the pictures and be responsible for how the programme looks as the director. You can sum it up simply in this phrase, 'the producer runs the shop and the director dresses the window.'

You also have to decide what kind of director you want to be – one that directs a small crew and manages a small team; a maverick film maker working almost singlehandedly, concentrating on single subjects and shooting the entire thing yourself or whether you want to direct many cameras as a multi-camera studio director.

Production versus development

Some researchers and producers alternate between working in production – actually working a show and making it happen and working in development – devising, creating and coming up with new programme ideas.

Although some researchers decide to become specialists in development, rarely working in production, it is invaluable to do both as it's useful to understand how a production works – whether working in a studio or on location with contributors and presenters.

'Only by producing programmes and making programmes can you work in development,' says Kate Phillips. 'Because you need to know how a programme is put together.'

By working in production you know what you can and can't achieve within the constraints of a programme's finances. 'People come up with ideas in meetings and I'm the one that has to squash them a bit and say, "Realistically, could you do this on £20K? You're talking about blowing your budget in the first five minutes because you want four helicopters dropping things". You need to have the production experience to be able to work in development.'

When do you need to decide where to specialise?

You can make this decision at any time but you really need to have made up your mind by AP level so you can focus your energies and ambition on getting on the right shows at the right companies.

'Figure out the kind of programmes you're interested in making,' says Matt Born. 'Treat your career seriously: keep your eye on your goal and take jobs that are more likely to help you get there rather than just the first thing that comes along, with the caveat that you've got to eat so that's not always possible.'

You must decide what you what you want to do – do you want to be a producer or director? A producing role involves writing, devising and producing content whereas the directing role is more visual – directing cameramen on location, in the gallery of a TV studio or self shooting the material and then cutting it in the edit. Do you want to become a self shooting producer/director or work in a studio?

'I was very focused on the type of television I wanted to make and did everything I could to make sure I worked in those areas,' says PD Nick Holt. 'I began as a researcher, then became an AP and eventually a PD.'

PROFILE

Conrad Green, executive producer, *Dancing with the Stars*
Conrad Green is a British multi-award winning executive producer who has been working in the United States since 2003 executive producing the hit show *Dancing With the Stars* and on *American Idol.*

Dancing With the Stars has been twice Emmy nominated in the Reality Competition category, and won many awards including the People's Choice Award in January 2007.

In the UK he was the series editor of the first series of *Big Brother* which has won numerous awards including a BAFTA for Innovation as well as a Royal Television Society Programme Award in 2000 and a Best Production of the Year from Broadcast in 2000. He series produced *Popstars* (ITV) which won many awards including a Silver Rose at the Montreux International Television Festival in 2001.

London born Conrad got his first break at Wall to Wall Television in 1994 and lives in Los Angeles with his wife and three children.

Programme credits
March 2005–at the time of writing (spring '09)
Executive producer, *Dancing With the Stars* Seasons 1–7, BBC Worldwide Productions/ABC Network

January 2003–January 2005
Executive producer, 19 Entertainment
Co-executive producer, *American Idol*, Season 4, Fox Network
Co-executive producer, *American Idol, Life on the Road*, Fox Network
Co-executive producer, *World Idol*, Fox Network (US), ITV (UK), RTL (GER) etc.
Co-executive producer, *All American Girl*, ABC Network

October 2001–January 2003
Head of New Entertainment Development, BBC TV London
Executive producer, *The Murder Game*, BBC1
Executive producer, *Diners*, BBC2 & 3
Executive producer, *Celebdaq*, BBC 3

May–August 2001
Executive producer, *Big Brother*, C4/Endemol

October 2000–May 2001
Series producer, *Popstars*, ITV/LWT

April 2000–September 2000
Series editor, *Big Brother*, C4/Endemol

March 1999–April 2000 Producer (series) at LWT/ITV
Pride of Britain Awards 2000, ITV
Tarrant on TV, ITV
Guinness World Records, ITV

December–February 1999
Producer/Director, *Miami Uncovered*, Sky One/LWT

July–October 1998
Producer/Director, *The Wedding Show*, Ginger Television/BBC1

February–July 1998
Producer (series), *Red Handed*, LWT Entertainment/ITV

January 1997–March 1998
Producer/Director, *Caribbean Uncovered*, LWT Factual/Sky One

August 1997–January 1998
Series producer, *Still in Bed With Me Dinner*, LWT Factual/ITV

May–August 1997
Producer/Director, *Ibiza Uncovered*, LWT Factual/Sky One

November 1996–April 1997
Producer, *The Show*, LWT Factual/C4

1994–96 Producer/Director at Wall to Wall Television/ITV
Director, *Big City*
Director, *Wired World*, Channel Four/Wall to Wall

How did you get into TV?
After graduating I tried everything I could to get in at entry level. My first job was cleaning and painting filing cabinets at Wall to Wall Television which led to a runner's job and eventual entry into the business.

What was your first break?
My first production job was as a junior researcher on *Big City*, a what's-on guide to London. It was a great entry level job as there was a very strong and experienced team who indulged my naivety and showed me what people actually did to make TV shows.

What attracted you to working in TV?
I loved watching TV and couldn't believe you could actually get paid to make it. It seemed impossibly glamorous so I thought I'd try to get into it, and work something else out if I couldn't.

What's the best thing about working in TV?
The variety, making something that people enjoy and want to talk about, and it's a good living.

What's the worst thing?
The hours and the job insecurity.

What's the best thing that's ever happened to you while working in TV?
Many moments with good friends and colleagues thinking 'I could never be here now, meeting these people and having this experience, if I did any other job'

The worst?
Agreeing to make a show I never believed in the concept for, and watching it fail. Life's too short and sometimes bad ideas are just bad ideas. It's important to believe in your own judgement.

Why did you decide to go and work in America?
I'd achieved pretty much everything I could imagine doing in the UK and the offer came. I've always believed you should take the most interesting and challenging offer you have when between jobs, and America is the biggest market in the world. Los Angeles is a great place to live and it's really refreshing to make TV somewhere so familiar but yet so different.

In the longer term I'd love to work in both the UK and the US, as there's so much to learn from both countries.

What is the best way to get ahead in television?
Work hard, use your brain and don't give up.

Assistant producer to series editor – what are the job roles?

What is an assistant producer?
After the researcher the next rung on the TV job ladder is assistant producer (AP). The role varies depending on what genre of show you're working on, but there are some common areas.

An AP should be able to:
- Write briefs
- Write scripts
- Write interviews
- Find facts
- Find locations
- Find and cast contributors
- Plot storylines
- Find and clear archive, stills and clips
- Take responsibility and manage runners and, researchers in finding and casting contributors
- Work alone on a strand or segment or VT of a show
- Specialise in one programme area

May also need to be able to:
- Self shoot
- Edit

'They'll be even quicker at finding information, have a good contact book for whatever their specialist area is. They should be completely versatile in that area by now,' says Richard Hopkins.

By the time someone has progressed to AP they will have chosen a genre of TV to specialise in, concentrating on working on a particular type of programme and have built up a track record in a particular role. For example, there are APs who specialise in contributor finding and casting, celebrity booking, development and shooting, and also in investigative research, finding and producing stories in a specialist area and working with a director or producer director.

'An AP often has to go to the producer or director and project storylines and you want to know that they know the beginning, middle and end of the story, rather than randomly saying something is happening,' says Julia Waring. 'They need to be able to take responsibility and give instructions as well as take them.'

'You use some of the skills that you had when you were a researcher but you use more of your judgement and analytical skills,' says AP Harjeet Chhokar who was a researcher for two years. 'As an AP you are starting to make more and more decisions, whether a contributor is any good, what you feel would work within a programme. These are really important skills that you pick up and I feel that I am using them more and more in the jobs I do. Also, you not only work closely with the researcher but also have to manage them as well. This can be quite difficult and needs to be handled very well to get the best out of the whole team. You also act as a go-between with your PD/SP, too.'

A self shooting AP is responsible for setting up shoots and filming on location, producing and managing the contributor and filming the whole thing on a light-weight handheld digital camera like a Z1 or PD150 or PD 170.

What's the difference between a shooting AP and shooting researcher?

The roles are very similar but the researcher has less editorial input and will probably be used more as a second camera for actuality and occasional GVs but the opportunity is there to show willing and to take on more editorial responsibility if you are an AP. 'It's what you make of it,' says Jon Crisp a proficient self-taught self shooter.

'If you're putting yourself forward as a shooting AP then it's important to know your stuff technically; how to use and troubleshoot the camera, frame shots, and shoot coherent sequences,' says Matt Born. 'A shooting AP will also have a better view of the bigger picture – the script and narrative. "It's also important you develop an editorial head, thinking about the story at all times. It's not good just shooting hours and hours of footage which has then got to be waded through, or switching off just as something interesting happens because you hadn't anticipated it or thought about the programme narrative."

What's the difference between a researcher and AP?

'It's the level of responsibility,' explains Richard Hopkins. 'In theory, an AP might be in charge of a researcher, it depends what the team structure is. An AP is more likely to pick up a camera and the responsibility that that entails, filming a piece that might appear on television. They would be more likely to direct content and know how to deal with talent (presents and contributors), and to write the last draft, if not the broadcast version of a script or an item at least.'

The tasks vary according to the type of programme, but generally an AP might have more say in the decisions that are made and how they affect the programme. 'As the AP you start to manage things more,' says Harjeet Chhokar. 'You may make decisions on casting or where to film. You may also be required to script and often go out with a camera and film secondary stuff. It's a great feeling, this level of responsibility, and when you're working in a great team, it all seems to flow more.

'You do have a little bit more responsibility,' says AP Dominic Crofts. 'You work closer to the producers. When you're a researcher the producer tells you what to do and you do it. If you're an AP you work close to the producer, you might get a chance to write scripts or you might get to go into an edit and cut together passages.'

An AP might have ownership of an area of a show, for example a strand or item, or interviews in a formatted documentary or reality show; they might even get a chance to get into the edit and assemble their material into a short VT film or sequence. 'Normally they would gain some editing skills of some sort, all round they would be a notch up,' says Richard Hopkins.

What key skills does an AP use in their job?

An AP is between a researcher and producer and is often a buffer and link between the two. They have to manage up and manage down and deal with contributors. 'You have to be a good people's person and have the ability not to panic, which can be quite difficult,' says Harjeet Chhokar. 'You need to keep a calm head and try to be reasonable, especially when things go wrong. Also, a thick skin helps. Every AP has worked on productions which have been a nightmare! It is important to try to just get on with the job in hand.'

An AP will often have the responsibility of managing the casting of contributors and keeping them onside. 'People in TV forget sometimes forget that for members of the public, appearing in a TV programme is the biggest thing that will ever happen to them and also one of the scariest. Therefore, you *must*, as an AP, have a great relationship with them and make sure they trust you completely and, for the two, three or four weeks you are filming that they can rely on you totally,' says Harjeet Chhokar.

Dominic Crofts earned his first AP credit after a difficult shoot on a pilot show with Russell Brand. 'It was the first AP job I did. I had to get at least three heroin addicts to Wales, really early in the morning. We got the train tickets. The producer and I talked about it and we came to the conclusion that heroin addicts weren't very reliable people and so we thought we'd at put at least one of them in a hotel the night before. We did that and the other two turned up but they were in a state and the other guy was having second thoughts. We spent hours and hours with them saying everything was going to be alright, and it was fine in the end. It's a case of managing people.'

Being an AP is a big step up from researching. As well as working within a team, sometimes, depending on the nature of the show, it can be an autonomous role where you get a chance to write, produce, direct and edit your own VT packages of short self-contained films, often without any training, help or input from anyone else and at short notice.

Being an AP is a great place to learn new skills and it is exciting. 'It's a lot more fun being an AP because you get to be creative,' says Dominic Crofts. 'When you're a researcher, well, you research, but when you're an AP you produce; you get to write a script, produce and direct it and then you get to cut the package yourself. You actually go into an edit and create a package with music and effects. And you get to deal with the higher elements in TV so your executive producer will come in and have a viewing with the commissioning editor whereas as a researcher you don't see the work until it goes out on TV.'

What does an edit AP do?
An edit assistant producer will assemble a sequence or part of a larger show that has usually been shot by someone else – usually a shooting team on location, from studio rushes on a multi-camera reality show, a formatted documentary or factual entertainment series. Often because the shows are reactive, topical or reality, there is no time for a PD to edit their own material when editing and shooting are happening at the same time.

What does an edit producer do?
Very similar to an edit assistant producer but with more experience and responsibility. Edit producers are sometimes and usually also producer/directors but may work as edit producers on specific

projects cutting sequences, entire shows or episodes of series. The role might involve adapting a British show for a market overseas – changing music, voice over and re-editing the show for time.

An edit producer should be able to:
- Create and plot storylines and a narrative from the raw shot material – the rushes
- Work out and plot a clear storyline and narrative arc
- Write an edit script
- Oversee the finding of supporting archive, stills and clips
- Take responsibility and work on a strand, segment or episode of a series

May also need to be able to:
- Self shoot
- Produce and direct

An edit producer will not have been on location or shot or directed the content but is responsible for creating the storyline and content of the show from the raw material (the rushes). 'An edit producer comes in at the point of the programme going into the edit, and works with someone else's rushes (or a number of people's), crafting them into a structured episode with a strong narrative arc. Sometimes you're lucky enough to get viewing days before going into the edit or access to logs of the rushes. Occasionally you have to go in blind and work out your structure as you're going through the rushes with an editor,' says Louise Mason.

She explains, 'If you're working within an established series or a series with a format, you have to work within established boundaries and then try to give your programme its own identity; that will include working on storyline, focusing on your particular contributors, etc. It sounds obvious but every story needs a beginning, middle and end so everything needs to flow and reach a satisfactory conclusion to your story arc.'

What does a producer do?

A producer is between an assistant producer and series producer and is normally involved in a strand or part of a big studio show – either a whole episode or a specific part of a larger show.

A producer should be able to:
- Take responsibility for a one off programme, single show or an episode that is part of a series
- Take responsibility for the runners, researchers, APs and manage the creative production team
- Oversee the writing of briefs and the setting up of shoots
- Write scripts
- Write interviews
- Find facts
- Find locations
- Find and cast contributors and oversee a casting team
- Plot storylines
- Oversee the finding and check clearance archive, stills and clips
- Specialise in one programme area
- Edit

May also need to be able to:
- Self shoot

'*Strictly Come Dancing* would be divided up into the studio show and the VTs that go into the studio show, so a live producer would be working to the live series producer organising that particular studio show,' explains Richard Hopkins who was head of format entertainment at the BBC and oversaw the development of the series. The producer writes the script, comes up with games, items, ideas and guests and will also edit the show to time if it is pre-recorded.

Daily magazine shows like *Trisha*, *This Morning* or *The Paul O'Grady Show* usually have five day producers who each oversees their small team, normally a researcher and assistant producer, and are each responsible for one show a week.

Entertainment series producer Anna Blue has had no formal training and has learned everything on the job. 'As a producer you learn from your peers and through experience,' she says. 'That's why working on a daily show like *The Big Breakfast* or *Richard and Judy* can be a great education. You get a huge numbers of TV hours under your belt and working in a big team allows you to pick up skills from these around you.'

Depending on how the schedule is arranged on an entertainment/chat show, the producer might be doing five a week or one a week depending on the run, overseeing different APs and researchers responsible for different segments and items.

On a studio show the producer might also be in the gallery speaking to the presenter, feeding them questions for the interview and helping them to deliver content. They'll speak to the guests and brief the production team, as well as overseeing and managing their team.

Producers are more commonly found on entertainment shows and often a producer will become a series producer and by pass the producer/director role.

What's the difference between a series producer and a producer?

Series producer and producer roles vary dramatically from production to production and between genres. 'Generally, a series producer is required where a production is so large that multiple producers are required to look after different elements of the show. A series producer then takes on the umbrella role of making sure that all the separate elements of the show work together as a whole,' says Anna Blue.

What does a producer/director do?

A producer/director (PD) will normally work on location either directing crews or self shooting content to a script they've written themselves. They'll edit their material and on a bigger formatted factual entertainment show may be part of a team, or on an observational documentary they might be the sole PD.

A producer/director should be able to:

- Take responsibility for a one off programme, single show or an episode that is part of a series
- Plan, set up and run the shoot on location, producing and directing the content, contributors or presenter
- Take responsibility for and manage a small team which might include an AP, researcher and runner (depending on the size and type of production)
- Write and oversee the writing of briefs and the setting up of shoots
- Write scripts and interviews
- Find locations
- Plot storylines
- Oversee the researching of facts and contributors
- Oversee the finding and checking of clearance archive material, stills and clips
- Specialise in one programme area
- Edit the programme, segment or episode

May also need to be able to:

- Self shoot

A producer/director might be in charge of one episode of a programme; they will write the scripts, shoot and edit the shows, and deliver it from start to finish. They will then show the finished product to a chain of hierarchy. 'The series producer comes in to give an overview. Then the executive producer comes and gives an overview of the overview, and the commissioning editor gives an overview of the overview of the overview, which can be quite painful but it's a process quality control,' says Richard Hopkins.

A producer/director will plan the shoots and storyline, discuss the contributor's journey with the series producer and executive producer frequently and be sure that it is all on track. They will direct crews and manage their team on location, they'll also write shooting scripts – writing the narrative and working out what shots they are going to film and where. They'll write a script for the edit – a plan of what shots, sequences and events they are going to film and where they are going to put the commentary, pictures and action, write a voice over script, choosing music, effects and archive for their programme.

'PDing essentially involves joining a production after the bulk of the research or casting has been done,' says producer/director Louise Mason, who gained her first official PD credit in 2003. 'I can be handed a storyline to maintain across a series, as in *Shipwrecked*, or contributors whose journey make up a single episode of a series, *Dog Borstal*, and I then plan what I'm going to do. I find it's important to keep the edit in mind from the start; what am I going to need from every bit of filming to help make the show work in the edit suite? Depending on the show, I can also use music as a good springboard for character development from the offset.

'As I don't self-shoot, a deliberate decision I made a long time ago, I work with a DSR two-man crew, often cross shooting with an AP. I direct the camera, but use wherever possible great cameramen who help me to focus most of my time on producing and story lining. Once the shoot is complete, I take the programme through the edit to completion.'

Anna Keel is a trained self shooting single and multi-camera director. She says, being a PD involves, 'coming up with a narrative; having a vision in your head of how it will play out, pace, style, music and all; working out what needs to be filmed to capture that vision; leading a team to put that vision into practice whilst remembering to stay well within the parameters of what your exec and ultimately the commissioner has asked for. Once the filming is in the can you need to work out how to piece it all together again into the vision you initially dreamt up, although generally it will have evolved a great deal since then. Once you've made a plan, you go into an edit suite and work with an editor to create the finished programme.'

A producer/director has to be multi-skilled – as well as being able to direct (and nowadays shoot) write and edit they have to manage people and be incredibly focused and organised. 'As well as creativity, being a PD involves a huge amount of organisation, attention to detail, people management and diplomacy. You also need to keep reminding yourself that you are not an artist in a vacuum, you are producing a product for a paying client,' adds Anna.

'Ultimately a PD has overall responsibility for what is seen on the screen,' says Nick Holt, a self shooter whose credits include single documentaries *Storm Junkies*, *Phone Rage* and *90 Naps A Day*. 'In practice, it's about managing the relationship with your contributors and understanding how various situations might be used in the final programme. Once filming has finished you then go on to make further decisions in the edit.'

What's the difference between a self shooting producer director and self shooting assistant producer?

'Quite a lot really,' says Nick Holt. 'You're responsible for shooting and sound, in the majority of cases, as well as directing what's happening in front of you. You've got to concentrate on both the technical and editorial aspects simultaneously. But with time this becomes less daunting.'

A self shooting producer will film the documentary or series themselves, using a handheld camera while often managing a small team of a self shooting assistant producer and/or researcher with perhaps a runner to help.

Jon Crisp says, 'As a PD, I have much more responsibility in writing shoot scripts, working out the narrative and having a clear idea of where the film is going and where I want to take it. The shooting and the edit are up to me. The biggest lesson I've learnt is to take a good step back from it during filming and during the edit so that I can look for the unravelling narrative rather than burying my head in it and losing track of its direction.'

He adds, 'You take on much more responsibility and have the opportunity to put your stamp on a film and grab it and take it where you want to go with it. I love it!'

I work either with full camera crews or as a self shooter. The difference is that as a self shooter you have ten times more to think about on location, and can't devote as much time to your contributors. It's far more important to have a good AP when you are self shooting for this reason. Working with a good crew who you can bounce ideas off is very rewarding, but self shooting is also a great buzz. It means that you can shoot more easily in situations where a crew might not be appropriate.

What does a self-shooting PD do?

'You tend not to have a sound person and cameraman,' says Nick Holt, who does the camera and sound himself. 'That's not say I never work with them. It's just that I tend to work with them less than most PDs might. Other than that, the job is pretty much the same.'

It's a demanding and difficult job that requires huge skill. 'You do everything that a PD does, as well as operating a camera and recording sound,' says Anna Keel who did BBC production and camera operating training. 'Camera directing is like driving a car. It involves keeping abreast of many things at once in order to operate the machine: framing, focus, exposure, colour balance, audio levels, all at the same time as directing and getting a good story out of what you are filming.'

Good training is available but most self shooting PDs learn on the job. Nick Holt learned to self shoot when he was an AP, learning from the directors he worked with. 'Some directors were very good at giving me a great deal of leeway with shooting. Everyone starts off with little experience and it's just a case of building on that gradually,' he says.

He supplemented this with some training. 'I've been very keen on photography for many years but I began filming seriously after a DV course with Urban Fox. It's run by Christina Fox, an ex-camerawomen who's fantastic at demystifying some very technical equipment.'

How do you go from being a self shooting AP to self shooting PD?

A good way of learning your trade, getting good credits and earning promotion is to work with an established and well-respected self shooting PD and work under their guidance and tutelage.

'I was a self shooting AP and gradually started to shoot more and more of the films I worked on,' says Nick Holt who became a self shooting PD in 2006. 'Eventually I was given the chance to direct by Tina Flintoff at North One TV on *Guys and Dolls*.'

But making the leap from shooting AP to PD is one of the hardest career hurdles to overcome in TV. The best way to overcome it is to get into the edit. 'My first official PD credit was in 2003,' says Louise Mason, 'but I had been running edits single-handedly since 2000. I had directed bits and pieces throughout my time as an AP too.'

She advises, 'Get in the edit as soon as humanly possible. I ran the edit as a producer for many programmes before I went out alone as a producer/director and it was invaluable experience. Watch and learn on every shoot you're on. There's probably a very good reason why the producer/director is doing it the way they are, and that's a great way of storing up ways and means of successfully capturing a sequence. There's no right or wrong way, different producer/directors do it differently, and the more variations you're aware of, the better informed you'll be to make a choice when it's your turn.'

A company that you've worked with before and have a good relationship with is more likely to give you a break and the chance to get edit experience. 'Get in with a company as an AP or DV director and, once they know and trust you, push for the chance to edit,' advises Matt Born. 'Be realistic: this does sometimes mean compromising on rate or taking a job that doesn't particularly light your fire. But the nature of the industry these days is that most employers are incredibly risk-averse and won't give someone they don't know a break as a PD, even if they're the most fantastic and experienced AP/DV director.'

'There's often a Catch-22 in making the leap from assistant producer to producer director, in that you are unlikely to get offered your first directing break unless you already have some directing experience,' says PD Anna Keel. 'The key is to find ways around this situation. One way is to put yourself forward for any bite-sized directing jobs either within a programme you are APing on, or as part of a magazine-style programme that needs short inserts. I ended up doing a bit of both.'

A self-starter as well as self shooter, Anna took the initiative to help her make the leap from AP to PD. 'My first real directing break came through a self-generated project that I managed to sell to BBC World,' she says. 'I'd heard they were looking for short, international news stories for their Travel and Current Affairs strand *Fast Track*, so I found a story in Vietnam and ran it by them. Because I was an unknown, I offered to work for nothing on the grounds that they pay me if they liked the finished product. They agreed, so I made a five minute film as a one-woman-band using a borrowed camera kit and two free days in an edit suite. It paid off. The film was broadcast in 200 countries and later again as the strand's "Pick of the Year" for 2004. After this I was offered a job as a directing AP with "some editing responsibilities". I ended up taking six of the half-hour programmes in the series through the edit, so rather than giving me an AP credit at the end, I left with my very first six PD credits.'

What does a multi-camera studio director do?

A studio director works in the gallery of a studio, directing the cameras – on a studio show this can be up to six cameras – and the vision mixer who will cut the shots the director calls. During the show the studio director guides the presenter and the crew. They cue the cameras, set up the shots, write the camera scripts, run the rehearsals and then basically take control of the technical side during the record of a live show. A multi-camera director is charge of everything you see on the screen.

A studio director should be able to:
- Run the gallery of a studio show working out the camera shots with up to six cameras
- Write camera scripts and allocate roles and shots to the different cameras
- Direct the presenter and studio crew – cueing lights, grams (sound effects)
- Control the technical side of the studio

May also need to be able to:
- Have PSC (single camera) directing experience
- Have location directing experience
- Be able to self shoot

'Studio directing is much more technical than directing on location as a PD. As a studio director you sit between the vision mixer and PA in the the gallery. You will have up to seven cameras at your disposal, all manned by operators on talk-back who can tighten and widen their shots or reposition completely whenever you give the command,' says studio director Charles Martin who has worked as a producer/director.

'Lighting, graphics, sound, auto cue, captions, and music can be called up and fired in from the comfort of your gallery chair. More often than not in studio, the action you are covering is being staged for the camera; consequently elaborate rehearsals will have taken place to establish the best angles and lighting effects to achieve the desired results. Ultimately, most studio shows are filmed as live, as opposed to making a piece of a programme; as the director you are responsible for seeing the whole of its technical delivery,' he explains.

What does a series producer do?
Almost at the top of the editorial production ladder, but not quite. 'The series producer is the lynch pin of the whole production,' says Richard Hopkins, who series produced the first live eviction shows for *Big Brother* in series 1. 'The executive producer dips in and out of

one production more, because usually they are looking after more than one production at once, unless it's a massive show in which case they look at various elements of production; a series producer is totally focused on one particular production.'

A series producer should be able to:
- Take overall responsibility for the content, look, style and casting part of a series
- Hire the team
- Commission the graphics, titles and music
- Take responsibility but delegate tasks to the creative production team runners, researchers, APs, producers, producer/directors and writers
- Hire and manage the onscreen talent – presenters and experts
- Oversee the writing of briefs, interviews and the setting up of shoots
- Write scripts
- Write interviews
- Oversee and have the final say in the choice of locations, contributors, presenters and style of the show
- Plot storylines, narrative arcs
- Oversee the finding and checking clearance of archive material, stills and clips
- Specialise in one programme area
- Take responsibility for the budget and make sure the programme comes in on budget
- Ensure the programme sticks to the schedule for pre-production, filming and editing
- Oversee and approve the final offline and online edits
- Attend the sound dubs of the show and sign them off
- Liaise and negotiate with the commissioning editor and executive producer on the content
- Ensure the delivery of the completed programme according to the schedule

May also need to be able to:
• Self shoot

The series producer takes responsibility for everything from managing the team, structuring, formatting, and writing the content and narrative of the show to the contributors, hiring and managing and producing the presenter, defining the style of the show and how it looks, and making sure it is delivered to the dates on the schedule on time and to budget. The series producer will oversee shoots, the edits and sound dubs, overseeing the delivery of the final product to the broadcaster.

'In very simple terms you are in sole charge of all the editorial and budgetary aspects of the production,' says Anna Blue who has been a series producer since 2003. 'The executive producer and the commissioner will probably have multiple shows to look after and they are relying on you to bring them a production they can happily sign off.'

It's a highly pressurised job with many elements – none of which you really learn at a lower level. You have to rely on common sense, judgement, intuition and the knowledge you've gained working on different elements of a show as an AP or researcher, hiring a good team you can rely on and creatives you trust. But ultimately you have to be sure of your own judgement and vision and have nerves of steel.

'In entertainment, and especially live TV which is my particular field, you need to be able to remain calm under pressure,' says Anna Blue.

It's a varied job with multiple roles requiring different skills that have to be learned on the job with no formal training. You need to be organised, creative, a quick and lateral thinker and good at managing people.

'In the early stages of a production the series producer will be responsible for hiring the rest of the production team and finalising the budget. From that point onwards they are in charge of the production as a whole, reporting to the executive producer and to the commissioner at the channel,' says Anna Blue.

'Their role is then to oversee all the editorial content of the show including the commissioning of the set, graphics, title music etc.

They must also organise the team structure, assigning roles to producers, APs and researchers. The series producer is responsible for signing off all studio and shooting scripts as well as VT inserts or show assemblies in the edit. During a studio show they will most likely run the gallery, communicating how they want the show to look to the director, making decisions about content and running order during the record or TX, and producing the presenters over talk-back.'

It's a difficult role because you have to oversee the entire production process when you might only have seen aspects of it until this point. And even if you are relatively inexperienced, you need to be decisive and take risks.

'You also need to be able to make decisions on a daily basis,' says Anna Blue. 'As a series producer you need to be good at management of time, resources and above all people. And whichever editorial position you might find yourself in, from runner to exec, you will need to be eternally creative.'

As well as managing the team, the budget and the schedule, you're also responsible for the content of the show, which despite the pressures and distractions of everything else has to be your main focus.

'You are the person who should always keep the big editorial picture in mind,' says Conrad Green. 'What the show is resides with you, and sometimes it is better to make any call than be indecisive and let the situation overwhelm you. You need to learn to understand and have an opinion about some of the "duller" aspects of show-running like budgets, schedules, contracts, channel relations, personnel development; all of which demonstrate that you really are in control of the show and will get you the respect you need to get people working well for you.'

You must also have a clear, creative vision for the programme. 'I like to shape the show editorially, that's the challenge,' says Alex Marengo. 'You need to have a good relationship with your exec producer if they are hands on and if they are not, the commissioning editor.'

It is a difficult and pivotal role as the series producer has to manage up and down, dealing with different personalities, getting the best from their teams, delivering their vision of the show and making sure it meets the expectations of the executive producer and commissioning editor.

'A lot of it is understanding and developing the idea itself and having the editing experience of making a show work. You need to understand what the commissioner is looking for. You have a variety of relationships: with the EP who is a full time collaborator, or if they are hands off, having more contact with your commissioning editor. You need to be in constant contact with them, being really honest about changes, delays and keeping on track with the schedule. If you leave it to the last minute it's not good. It's a big editorial role,' says Alex Marengo.

Which means you have to be adept at solving problems under pressure and incredibly pernickety about the detail.

'You have to make sure that people are telling you if there are likely to be delays or possible overspends,' says Alex Marengo. 'Make sure that shooting and post-production are properly sorted. There are lots of big money pits you can fall into. You have to be a bit of an anorak!'

'The series producer should be totally obsessive about every aspect of the running order, be extremely conscious of the budget for any shoots they're organising, managing the team to maximise their effectiveness from top to bottom,' says Richard Hopkins. 'It's all the man and management skills you possess that come into play. The series producer has to work out how to make the show stand out, be creative, make the format work, make the subject of the format shine through so it really comes across as a real show that people will feel resonates with their life. 'All these factors are the series producer's job. Quite often you might be given a show that has been pitched in one way by a development team, but actually you can't make it in that way.'

This is not uncommon, my first series producing job was working on a series that had been sold on the back of a clever idea for a first round of a game show. The show brief was literally a paragraph long. I had to create, format and work out what the show was and the remaining three rounds were while the show was in production. I had just six weeks to pull it together when we hit our first filming block and I had two teams on location shooting it across Europe, Japan and America.

I went from being one of five live producers on *The Big Breakfast* managing a small team within a massive team to series producing the first run of a 13-part series and managing a big team of producer/directors, APs and researchers that I hadn't met before. Although most of the team were great, there were some who were difficult to manage – particularly the male PDs who weren't happy to take direction from a first time female series producer, and also awkward personalities who didn't want to work with each other. I had to broker relationships, mend friendships and play 'Mum' to keep everyone on side and the production together – all without any training or advice!

'Series producers really need training to manage people,' says Alex Marengo, who has worked in TV since 1990. 'I won't pretend to be the best manager. It's taken me a while to learn the real basics. It's all on the job training. You make mistakes. It's how this industry works.'

Often nothing prepares you for going from producer to a series producer as the leap is so great, there is such a skills gap and the increased responsibilities are enormous. Because there is little training and because many producer/directors are reluctant to become series producers because of the pressure and responsibility, there is a real lack of them in the industry; a fact that Skillset recognises. In 2008 DV Talent won funding to put on the first industry training scheme for series producers. Because budgets are squeezed, production schedules have got tighter and shorter – putting extra pressure on the production team making the programme. 'You are asking people to do virtually impossible jobs, there's never enough money or enough time. You are always asking your team to do more than someone in a nine to five job would be expected to deliver,' says Alex Marengo.

'You need really strong people management skills; to understand when to cuddle, to shout and to be diplomatic,' says Conrad Green.

It's also about making sure you put together the best production team to deliver the programme – and a reason why people choose people they trust and who have the right track record. 'It is all to do with choosing people who can deliver and have the best attitude,' explains Alex. 'If you choose people you trust them, and you give them the responsibility and the opportunity to learn. As long as you've got a dialogue and can work out problems. People want to make the best possible film or item. You can't put up with laziness. It's one thing to pull your finger out and do more than is asked for but if you don't do what you should be doing you won't last very long.'

What skills do you need to be a good series producer?

As well as managing relationships to ensure the production process runs smoothly, you need to be able to think laterally, write well, have good ideas, be able to produce, direct and edit well.

Writing is a key skill that underpins everything, as you will be responsible for writing and overseeing the script, which is essentially the format, the vision and the content of the show. You need to be able to write pre-shoot and edit scripts, commentary for a show but also to communicate with your team in writing, via emails, etc.

'It's vital that you can write,' says Alex Marengo. 'Put in writing what it is you expect of the team. You will very often write treatments, outlines, and step into the edit to help if someone is too close to their work, to help them. You have to be clear what is needed, about what needs to be said. You have to be in charge of a lot of detail and make sure it doesn't slip and as well as having an overview of the budget, schedule and post production.'

I've series produced six different productions; I worked around the clock and sometimes for seven days a week. A series is almost like a child you give birth to, get on its feet and then deliver to someone else. You are under pressure to deliver the best product and sometimes without the budget you need or the best team. There is always a weak link! Things never run smoothly so you constantly have to think of new ways to make things work. All with a smile on your face and a cool head!

You can only really relax when the show is finished and is on air, which is an incredibly awesome and satisfying feeling. Alex Marengo has been series producing since 2004 and he says, 'I've had creative input, I've done pure series producing, finding people, working on scripts, casting, shaping the series with the exec. It's very satisfying, to come up with something and feel quite proud of the finished product.'

What's does a series editor do?

The position above a series producer but below an executive producer, a series editor (SE) is responsible for the overall editorial tone, content and structure of a series – with key emphasis on post production. They are involved with all stages of production throughout the making of the series and work closely with the series producers and directors. They will usually report to the executive producer who will maintain the relationship with the broadcaster.

A series editor should be able to:
- Write the script and devise the programme format and narrative style
- Hold the creative reins and ensure quality control
- Take responsibility for the content, style and series structure
- Run and oversee the edits of a show, managing the producer/ directors, edit producers, and edit APs

Sue Dulay is a series editor at Ricochet Films. She explains, 'At the start of a new series, the series editor will draw up a format script template for the team to work to. As a formatted series requires a tightly structured story to be told in a similar way within every programme, all potential stories are assessed by the producers and then discussed with the series editor to ensure all format points can be fulfilled in a creative and entertaining way to give variety within the frame. Key pieces to camera, discussion sequences and format points are written up, scripted and approved by the series editor before filming. Once all material is shot, the series editor oversees the construction of the programmes in the edits, often running many suites at once with several edit producers working on a couple of shows each.'

How do I choose which genre of programming to work in?

You must carefully decide which genre and programme area you want to work in as the skills and knowledge required are different. Factual is different to entertainment and features, documentaries different to reality. It is best to choose the genre best suited to your sensibility, personality and knowledge, and follow that chosen route. 'You mustn't lie about what you're interested in because you won't be able to give your heart to a particular type of programme. Go by companies that are appropriate, find out about them,' says Julia Waring of RDF.

Most companies have different brands and are known for making particular types of programmes. For example, Optomen make cooking shows – they found and developed programmes with Gordon Ramsay and Jamie Oliver; Princess Productions and Monkey Kingdom make entertainment shows; At It Productions make music and youth shows; Darlow Smithson has a pedigree in making big dramatic specialist factual series on science and history.

Most people know where their sensibilities lie – even before they start working in TV, so it's really about being true to yourself and being honest about your talents and interest. Katie Rawcliffe knew from a young age she wanted to work on entertainment shows. 'I really liked dancing when I was younger and I wanted to do something that was creative, either dancing or music. All the things I'm working on now, the entertainment shows such as *Dancing on Ice*, I really enjoy. It goes back to that really.'

Similarly, John Adams knew he wanted to work in entertainment TV because he loved watching *Noel's House Party* and the Barrymore shows. He says, 'I got to go to *Noel's House Party* a few times. I was shown the gallery and decided that that was what I wanted to do, live entertainment shows. *Noel's House Party* then used to get around 13 million viewers. I'm just waiting for *Noel's House Party* to get re-commissioned so I can go and work on it!'

However, Conrad Green never focused on one particular genre of programme making. He says, 'I never really have. I've made documentaries, magazine shows, factual entertainment, clip shows, hidden camera shows, award shows, straight reality and old–fashioned entertainment shows.' This is possible if, like Green, you work at a big independent company like ITV Productions which make programmes in many different genres.

'It is an asset to have diverse experience in the modern TV world,' says Green. But it is important to focus on one genre and plough a particular furrow. Play to your interests. Work out where your strengths lie and pursue the jobs that can best make use of your skills and experience. Once you've established a genre it is a good idea to pursue it and become an expert in that area.

Alex Marengo is an established specialist factual series producer. Although he started out as an entertainment researcher he decided to specialise in drama documentaries. 'I always wanted to work in drama and a route into drama was to do drama documentaries first. These were being made by BBC Music and Arts. I got a job as a researcher at the BBC and stayed there for years becoming a producer of magazine shows. I got to make twenty-minute films. It was terribly seductive. I filmed all over the world; it was great. So I became sidetracked from drama and began to make drama docs. I tried the *EastEnders/The Bill* route but I like real stories, human interest stories, it doesn't matter whether in history or culture.'

When I am crewing up for a project I look at someone's CV to see if they have similar and comparative credits and experience in a certain field. I want to know that they are familiar with making a particular genre of programme. If I am hiring for a specialist factual show on science I want someone with a proven track record in this area and good credits on similar shows.

Similarly, if I am looking for a factual producer/director to make a documentary for Channel 4 at 9pm, I want to see that they have experience and a good pedigree of making this type of show from the quality of programmes they have recently made – that they have made good, high-profile documentaries for Channel 4 that have gained a big audience or critical acclaim, or hopefully both.

Like most broadcasters, Channel 4 has an unofficial, unwritten list of preferred producers – PDs, APs, SPs and EPs – who are hot, who it rates and wants to work on its shows.

Commissioning editors have the final say as to who makes their programmes. Shows are commissioned on back of the off-screen talent and the good name and reputation of the production companies making them. Just as individuals have reputations so do companies. The truth about commissioning is that several companies might have the same idea at once. Who gets the commission is based on serendipity, personal relationships and the attachment of key and prized programme makers – whether PDs, series producers or execs. Whatever the commitment to growing new talent, commissioning editors don't like taking risks and will return to the names they know and trust to deliver their programmes to the highest standard, even more so in times of recession.

Once someone has had a hit their phone will not stop ringing. This top twenty five per cent cream of talent will be constantly in demand and constantly in work. They'll name their price and cherry pick their projects while the rest of the jobbing freelancers chase the rest of the work available to try to stay employed.

You have to be strategic and get in with the right companies and on to the right shows with the right people. TV is about making hits, achieving viewing figures and critical acclaim. Broadcasters and production companies want the best – the hit makers.

Broadcasters are under pressure to deliver audiences and ratings and so will return to the successful production companies to ensure that they get the hits they need, rather than take a chance on someone less well known or successful, no matter how talented.

Misdemeanours are usually quickly forgotten if you're a hit maker. The programme maker Stephen Lambert resigned from RDF for editing the footage of the Queen that caused such a national scandal. He was the format creator and hit maker behind the indie's biggest and most genre breaking hits – *Wife Swap, Faking It* and *Secret Millionaire*. So it's no surprise that when he set up his own indie, Studio Lambert, that it had a raft of new commissions from Channel 4 within weeks of opening.

As you progress to senior positions it's good to target companies where the producers have a sensibility or style that appeals to you and are working within a company that is dynamic and growing.

'I have usually chosen my jobs because there is PD or SP who I have always wanted to work with or because I really admire the company or programme or they have great people working there,' says Harjeet Chhokar.

How easy is it to work abroad?

The UK has a well-earned reputation as a TV innovator, selling successful formats like *Who Wants to Be A Millionaire? Wife Swap, Supernanny,* and *The Weakest Link* around the world. 'Something like over fifty per cent of formats today come from the United Kingdom, which is amazing if you think about how small we are; we're like the empire of television,' says Kate Phillips of BBC Entertainment who has consulted on the BBC's shows in the US. The BBC had *Strictly Come Dancing* and now the US version *Dancing with the Stars* is doing well internationally. In the old days that just didn't happen. It was always American bought, when we were growing up it was all American imports.'

The success of *Who Wants To Be A Millionaire?* in the US reversed the trend. 'Suddenly America is looking to Britain; we had *Pop Stars* and *Pop Idol,* and then America did *American Idol,*' adds Phillips. 'I think the last *American Idol* got sixty million votes in the final, more than any US election.'

The UK is now a key creator and generator of ideas. Conrad Green is *Dancing with the Stars'* executive producer who has overseen the programme's enormous US success. He says, 'In many ways the UK is ahead because the public service remit means that UK broadcasters are expected to innovate.'

Working on a hit show in the UK the format of which was created here and the idea has been sold to other territories offers the opportunity of exporting your knowledge and working abroad. 'I did a lot of consultancy because *Dancing on Ice* sold internationally,' says Katie Rawcliffe of ITV Productions who series produced the hit show's first run.

'I've consulted in Australia, Spain and Russia. A lot of it is done beforehand by email and on telephone conference calls. Then, depending on where they are in their production period you go over at the beginning and sometimes when they're actually making the programme. I met the producers and they asked how we did it and I gave them advice. It's good fun. It's also interesting to see how other countries make television programmes because inevitably it's different. It's a great experience, a good thing to do.'

British TV producers with hits on their CVs have a good calling card to work abroad and achieve great success by taking their production expertise and experience overseas. Conrad Green has been working in America since 2003 and has reaped huge financial rewards. He first went to the States to co-executive produce *American Idol.* He liaised with production companies in ten countries to produce it and has now completed six seasons of *Dancing With The Stars* in an incredibly tough market.

'TV in the US seems much more hard-nosed than in the UK,' he says. Without UK public service justifications like innovation and quality, the only barometer for success is ratings, and in a very narrow demographic. The overnights for your opening show in the US can be a really, really brutal eye-opener.' In the US failing shows can be cancelled mid-run, something that happens in the UK but is not as common.

Green says TV productions in the US and UK are the same. 'The skills and mechanics for the shows are similar, it's just the job titles which are different. In the US a series producer is known as a show runner, but the expertise required is the same. In a show runner everyone wants someone to bring good editorial and management judgement, confidence, security and success. That's the same in both countries,' he explains. 'You need a bit more political experience to navigate the US network market and you need to be a bit more circumspect about how you deal with people.'

Green maintains that the skills required to work on shows in the US are no different to those in the UK. 'Essentially they're the same,' he says. 'Generally, storytelling needs to be more punchy, shows move faster and production values are very high on US network TV.'

Fellow Brit Richard Drew, who has been working in the US since 2005, agrees. 'The skills behind making a show are the same wherever you go. The key issue is local knowledge, of course. That's why, when British companies try to bring out British production staff, it generally doesn't work. A British production manager is just going to be lost in the US. How can you run a production when you don't know permit laws, tax issues, suppliers?'

Richard set up the US office of a British independent production company from scratch. 'It's been a real rollercoaster,' he says. 'I came out to set up Zig Zag's US office but unlike a lot of other British production companies we didn't have a big series or commission to get us started, we just had a couple of one-off pilots. So I was running these shows and trying to generate more work at the same time and we were very new and not known in the market, so it was difficult.'

The way to survive and conquer the US market is to establish a presence by networking and to understand the way the market works. 'One of the key challenges is that you really have to be on the ground in New York or LA to know the market. You have to be watching TV constantly, taking meetings, building relationships with networks and that takes time. The first eighteen months were very difficult,' he says. 'But now I'm very well known in the New York market and have these relationships and the trust of many of the networks and that makes a huge difference. They can see that I understand the US market and the sensibilities of American TV. And I've sold my own shows now and that's crucial.'

Drew has executive produced several different shows for different American broadcasters in the US, hiring American staff. He says that although the job roles are similar, the whole staff structure is different. 'There is much less of a hierarchy, people don't go from runner to researcher to AP. It's PA straight to AP or even producer! And there are so many types of producers: field producers, segment producers, story producers, supervising producers . . .'

I've looked for and recruited US talent for British companies making shows for American broadcasters. It's a difficult task as the job titles are different and when you scratch beneath the surface, the roles that an AP or producer has performed are different to practitioners in the UK. Because of the different number of roles in the US, producers are less skilled and less able to multi-task, are unable to perform the different roles expected of a British producer.

'The roles are a lot more segregated, especially when it comes to producers. It's harder to find people who are multi-skilled,' says Richard Drew. 'In the UK, if you're a producer, the chances are you can find contributors, edit, self shoot, write scripts, etc, because you've come up through the ranks and picked up skills along the way. That's not really the case in America. It's because most producers in the UK have spent a few years at researcher and AP level. But you can find many people who call themselves producers in the States and yet have fewer than a couple of years TV experience under their belts. There's also a cultural difference. Brits will undersell themselves for a job. Americans will oversell themselves, especially in LA. And often they just don't have the experience or skills to match their words. That said, there's just as many good people in the US as in the UK, they just sell themselves differently.

'My rule of thumb, and this goes for both Britain and America, is that ten percent of people in TV are amazing, eighty percent are OK,

to varying degrees, and the other ten percent are absolutely awful.'
Richard has set up his own company to help UK companies establish
a base and find talent in the US.

'It was incredibly hard when I first moved to the US but now I've
been here for a few years I know who the good people are and how
to find them. That's why it's so hard for British indies when they first
set up shop in the US, they're literally starting from scratch!'

PROFILE

**Richard Drew, executive producer and managing director,
Savannah Media**

Richard is an experienced executive
producer who after 10 years of working
in the UK TV industry, relocated to New
York and has since become a well-known
producer in the American TV market.

Richard began his career as a
runner in the mid-90s, working on the
children's TV show *Scratchy and Co.*
He worked his way up through the ranks of the British TV
industry, working on a variety of series including *The Big Breakfast.*

In 2000 Richard worked as an AP on the first series of *Big Brother*
and has since become highly experienced in the reality TV genre
having worked on *Survivor, The Club, Fame Academy* and *Fashanu's
Football Challenge.*

In 2005 Richard relocated to the US to set up and run Zig Zag
New York. He built the US operation from scratch and generated
several US commissions for the company, most notably creating
and executive producing the music reality series *Redemption
Song* – the Fuse network's biggest commission to date.

In 2009 Richard started his own production company in
New York, Savannah Media, a traditional production company
which also acts as a consultant to UK companies looking to pitch
and establish a presence in the US TV market.

Programme credits

2005–at the time of writing (spring '09) Executive Producer Zig Zag New York

Redemption Song (Fuse)
Around the World In 80 Babes, Playboy Channel
Going For Broke, Lifetime
15 Films, A & E
Cool Camps, Travel Channel
You're Not Getting A Dime, TLC
Hampton Nannies, TLC
News of the Weird, TLC

2003–2005 Series Producer

Everybody, Channel 4
That's So Last Week, Zig Zag/five
Fashanu's Football Challenge, Zig Zag/Bravo

1995–2003 Producer

The Club, Carlton/ITV1
Fame Academy, Endemol/BBC
Survivor: Panama, Planet 24/ITV1
The Big Breakfast, Planet 24/Channel 4

Other production credits

The Dance Years, ITV1; *The Gay Team,* five; *Planet Pop,* Channel 4;
Big Brother, Channel 4; *Scratchy and Co,* ITV1; *It's Not Fair,* ITV1;
Now and Then, ITV1; *Massive,* ITV1.

What does your job involve?

The starting point for my job is ideas – brainstorming ideas, crafting ideas, meeting with talent to build ideas. As a company you have to be on the lookout for new shows and new opportunities constantly. I take regular pitch meetings with US networks and make sure we're constantly in touch and developing formats that meet their specific needs. Then when we sell a show I executive produce the production. Alongside this, I oversee the daily running of the company, the management of the office and hiring of all staff for productions.

We basically act as consultants for UK producers who don't have the resources or finances to set up a US office. We allow them to minimise the risks while also opening up a whole new market to pitch to.

How did you get your first job in TV?
I started as a runner in TV in 1995 and I basically worked my way up the ranks from a runner to a researcher to an AP to producer to series producer and to executive producer. I paid my dues and did my time in each role and worked my way up. You learn so much from each job you do and I'm really glad that I have a wide production experience to draw on based on all my years in the field, in edits, developing shows, etc.

Did you have a game plan?
Not really because I think it's really hard when you're first starting out in TV to have a game plan. I didn't say, 'By the time I'm 30 I want to do x, y and z.' For the first three or four years you just have to work. If you are given offers, you take them and work. You can be pickier later on but not in the early stages. My only game plan was to go the researcher, AP, producer route but there is no timescale as to how quickly you can do it.

How have you balanced staying employed and working on good shows?
There are some jobs that you do for the money and some that you do for the experience. When I finished the first *Big Brother* I was offered *Planet Pop*. I really didn't want to move up to Manchester for three months to do the show. But it was good experience. I was going to be directing; I'd be filming interviews and mini music video shoots and longer packages. Even though I'd directed before they had been very short packages, so I took it because I thought it would be really good to get more extensive directing and editing experience under my belt. And overall it was a good move for me and I learnt a lot from doing it.

There are definitely a couple of reality shows I've taken on more for the money, so they'd give me choices to do other stuff. But I'd say every show I've done I've at least had some interest in. I won't take a show that I know is going to be a complete nightmare or I'm going to hate from the start. I could never do *Top Gear* because I have absolutely zero interest in cars. But for many people out there, that show is a dream job! You shouldn't take on shows that you know you are absolutely going to hate.

I had an interview with a production company a few years ago. I'd worked with them previously and as a company found them quite difficult to work with. They cut a lot of corners production wise and the vibe of the company was very unfriendly. They told me about the show and I realised it was just going to be a production nightmare. They'd had so many people drop out of the show and they wanted me to start literally the next day. It just felt like a really troubled production. I really needed the job but I just felt in my gut that I really didn't want to do the show. I turned it down. Then a friend of mine got a job on the same show and confirmed that it was a complete nightmare!

What advice would you give to someone starting out?
Be persistent. It's all about knocking on doors, sending out CVs and doing everything you can to get your foot in the door. Everyone knows the statistics for getting into the industry are grimmer than ever but there are still jobs out there and someone is going to get them. I firmly believe that if you keep pushing and stay positive you will get those opportunities.

When you do get an opportunity make the very most of it. Show interest, be cheerful, stay late, work your socks off because that's what people will notice. They want someone who is likeable and reliable and shows a passion to get ahead. I can't describe how many times I've taken on runners and they've done an OK job but nothing exceptional, and then they wonder why they don't get the second or third job. Getting your foot in the door is only step one. You always have to prove yourself at every stage of the industry.

What skills do you have that have contributed to your success?
Honestly, I think it's largely about passion and enthusiasm. I am a good writer, have good ideas, good management and people skills but a lot of those skills have been honed on the job over the years. You learn by doing the job and that's why I'm glad I worked my way up through the ranks instead of being fast tracked.

But that enthusiasm is something you can't learn, you either have it or you don't. If you aren't willing to devote large chunks of your life to the job, stay late, work weekends, and do it all with a smile, then this industry isn't for you.

You've forged a successful career in the US – how have you done it?
I was given a great opportunity to come out to New York in 2005 and I took it and ran with it. I had always had a huge love of America and American culture so my knowledge of the US market was already pretty good. Being on the ground really enabled me to immerse myself in the TV scene. To put it simply, if you want to operate in the US you have to know the channels and you have to know the shows, and the only way to do that is to live here.

Zig Zag USA opened in a different way to other UK indies that have set up Stateside operations. Many of them – Lion, Leopard, Tiger Aspect – came out with established series already sold to the US so they had the safety net of an existing commission and the budget to build a proper infrastructure. I came out to the US completely on my own. For the first year ZZ NYC was me and a desk! We had a couple of pilots that got us off the ground and gave us some production work but everyone knows you don't make money off pilots, so it was really important to build up the business and fast!

I took a lot of meetings, worked on a lot of ideas and generally just got myself better known and the work started to come in.

There were a lot of ups and downs, and you inevitably make a lot of mistakes when operating in a brand new market with limited support. Many things are different in America and it takes a while to really get used to the culture.

That's one of the reasons I decided to open Savannah. I'm probably one of only a handful of Brits who have set up a successful US indie from scratch: I thought there was a real gap in the market – acting as a stepping stone for UK indies with no US presence – that we could cater for.

What's the best piece of advice you were given?
TV is all about 'fire fighting', someone once told me and it's absolutely true: it's one problem after another. A shoot will never go easily. You will always have budget issues. At some point a guest will pull out on you at the last minute, a camera will break before a crucial shot, a staff member will take another job and leave you in the lurch. You have to accept that and make contingency plans. Go in with your eyes open and try to cover as many eventualities as you can.

I also once read in a book the phrase 'just pick up the phone' and those five words are so powerful. There are always hard calls you have to make, no one wants to hear 'No' but just pick up the phone and make the call. Whether it's calling a company for a job or looking for a contributor. That one call could change your life.

What have been your best moments working in TV so far?
It's always a great feeling when you deliver a show and the network loves it! Interviewing some of Hollywood's finest for *The Big Breakfast* was a real thrill as I'm a huge movie fan. The finales for *Big Brother* and *Fame Academy* were great nights, though obviously stressful behind the scenes!

What's the worst thing about working in TV?
The stress. It really is intensely pressured at times. There's never enough money or time and yet everyone has huge expectations!

There are always so many elements that can go wrong on a show so keeping your head and soldiering on can be tough. It can really take a toll on you at times.

What's the best way to get a job in television?
Make it your number one mission. I strongly believe that if you want something badly enough it will happen, it's just a question of time. That's why TV tends to be a young person's industry, you tend to have fewer distractions in your life at that stage and you have the energy and single mindedness to devote yourself to cracking the industry. So go for it!

Chapter Six
The future

What's the future for TV?

The modern media business is sometimes viewed as being at war – where the break neck growth of the internet is killing off 'old' media like TV, radio and newspapers. The value of internet advertising is set to overtake that of TV for the first time in 2009, according to a report by the Internet Advertising Bureau. But although these newer technologies are transforming the ways that media owners do business – like delivering TV over the internet – 'traditional' TV is here to stay. Consumers are not abandoning it. There is plenty of research to show that the internet has supplemented rather than replaced traditional TV consumption.

The television industry marketing body, Thinkbox, has research suggesting that households with digital television recorders, which allow consumers to time shift viewing, actually watch more television. The company has carried out a study with the Internet Advertising Bureau to show that people are often watching TV and surfing the internet at the same time.

TV is evolutionary and that in whatever format the content arrives as entertainment it will continue to mutate, adapt and be as popular as it has always been since it was invented.

'Just as the film industry and radio didn't die when TV came along, the traditional idea of telling a story over the course of half an hour and entertaining people in a format they can sit down and watch on a screen, what you and I think of as traditional TV programmes, that will survive,' says Fiona Chesterton, former director of television at Skillset.

TV in the UK has evolved from one corporation and two channels to a multi-channel and now multi-platform medium. Anyone wanting to sustain a career in TV has to be aware of how making and delivering a programme has changed and will continue to change in content and platform. It is no longer sustainable or profitable simply to deliver a show to a broadcaster.

'There's consensus that people will be seeing TV content as something which increasingly has a life or a form in ways other than just on your screen in your living room,' says Fiona Chesterton.

The holy grail is to produce formats and content that can service TV and also different mediums and markets. A format should be transferable and work in several different territories overseas, on other channels and other platforms like the internet and mobile phones, through spin off shows or edited highlights. The phrases '360-degree commissioning' and 'multi-platform' have become TV buzz words which reflect this new universal and rounded appeal.

'You hear the phrases, 360-degree and multi-platform content because if you have a television programme that works, it has another life completely outside the TV programme,' says Fiona Chesterton.

BBC3 controller and former head of E4, Danny Cohen, defined the phrase for an RTS lecture. '360-degree commissioning involves trying to commission content across all platforms at the same time . . . not adding on piecemeal later when you realise that you need to be multi-platform'.

New media expert Peter Cowley, managing director of digital media at Endemol UK, added that 360-degree commissioning meant 'fitting the most appropriate platform to your idea. It could be one platform or many.'

Makers of *Big Brother* Endemol were the first to successfully make a TV show accessible through different media on a grand scale through the revolutionary format which changed the face of broadcasting in the new millennium.

It created the culture of phone voting which effortlessly provides a huge amount of money with no need for costly production values and provided 24-hour low-cost content for other channels like E4 and More 4 through additional spin-off shows and through live streaming, also available on the internet.

But the reality show's pioneering use of multi-platform outlets wasn't planned. '*Big Brother* wasn't indicative of the 360-degree principle,' says Peter Bazalgette, the former chief executive of Endemol who made the show. 'It was a brilliant idea and serendipitously took off simultaneously in 1998/99. Actually quite literally last minute after its commission for broadcast.' Endemol in Holland, which first broadcast the show, latched onto the internet almost as an afterthought, to harness the new medium which was taking off at the same time.

Live events and stripped transmissions (where a programme is transmitted daily for a set period, like *Big Brother* or *I'm a Celebrity . . .*) are now common among all broadcasters and channels but the revolutionary series was the first to strip daily transmissions across C4 and show raw material. '*Big Brother* was the first show in the history of TV to show its rushes to the public, there were a lot of arguments about how unfairly edited it was, as it's the only show where the public has the right to have an opinion on that as they've seen the rushes,' says Peter Bazalgette who is now a consultant to Endemol.

Endemol has diversified its output even further by making the content it produces available on several different media and is now 'platform neutral'. Says chairman Tim Hincks, 'We are doing a part air, part interactive comedy show for BBC 3. Our ideas now are working their way onto other platforms because an idea is an idea. It's about making a digital story work without restructuring or changing it in any way because at the end of the day we are, when you strip away the layers, an ideas company. That is what we are. And most of the ideas are for television. But we're also doing shows on Bebo and we're doing shows on mobiles.'

I worked in development at CBBC Factual, developing factual entertainment formats for the channel and for BBC 1. When devising and writing formats the shows had to fulfil a number of criteria outlined on the commissioning forms: as well as being innovative, interesting, original and engaging to watch, they also had to work on several different platforms – the internet, on mobile phones, overseas – and satisfy audience research. We had different meetings with various BBC departments, including Interactive, to create spin-off ideas for the CBBC website, creating interactive games for 6–12 year olds, the channel's audience.

TV is aimed at an increasingly young, media-savvy and technologically aware audience; it is not enough to have a simple one-off idea nor is it cost effective for companies to make these ideas.

It's also staffed by young people – people in their 20s who are researching, in their 30s producing and as new entrants come into the industry they bring a different awareness of technology and view TV in a different way.

'People coming into the industry now, age 21, 25 or even 18, are the first generation who have grown up completely with digital media and are completely comfortable with it,' says Fiona Chesterton. 'Therefore, it is likely that when thinking about good ideas they will think about these in terms of how it can be consumed and not just as a static one-off TV show.'

Young people also lead the way in which television is physically viewed. The BBC's Kate Phillips worked in children's before taking up her post in entertainment. She says, 'It was actually a huge eye opener for me. Children have always known the internet and what we found is that they don't distinguish between the computer screen and the television screen. For them it's all just where they get their TV content from. Eventually in our houses there will be one box which will be the TV and the computer; it will be all in one.'

This also means they are less engaged – it's more laid back, low-attention viewing than high-quality appointment-to-view watching, which means programme makers and broadcasters have to work harder in attention-grabbing, talked-about television.

'When there were three channels, TV was very rare; now it's ubiquitous, anybody can shoot, edit and distribute their own content, so TV is no longer special,' says Peter Bazalgette. 'Though it does try to be. You have ten million people watching *X Factor* on Saturday night. It does have a whiff of being special.'

ITV, five and Channel 4 have responded to the changing world by creating digital channels, building websites with video on demand services and developing internet protocol television (IPTV). ITV has invested heavily in its digital future with the launch of ITV.com, the roll out of its ultra-local TV service, ITV Local, which relies on user-generated content, and the purchase of Friends Reunited.

With an expanding network of channels and endless choice, more channels are competing for fewer viewers. Programme makers and broadcasters have to work harder in attention-grabbing, talked-about television and there's less advertising revenue. As well as the traditional 30-second advertising spot, broadcasters are selling sponsorships and advertiser-funded programmes such as ITV's collaboration with Pedigree dog food for *Dog Rescue*.

'Advertiser-funded programmes are a bit of a misnomer, we call it branded content,' says Ben Barrett, commercial director at Zig Zag Productions. 'It's taken a surge in recent years because of digital media. Because of TiVo and Sky Plus viewers can fast forward so they are not consuming the ad content in the way they used to, so advertisers use advertiser-funded programmes to get their message across.' So, instead of just paying for ads during a break in a popular programme, manufacturing companies will entirely fund broadcasted programmes that will carry their brands and products, a practice that started in America in the 1950s.

'Soap operas were funded by companies in the States, hence the name. It's definitely happening here and will happen more and more. Branded content is a way for a company to get in front of the right consumer. You find a programme concept that marries itself naturally to the brand values of a certain product, so the brand is getting its message across not by an advert but through the editorial of the programme.'

Duracell entirely funded a science series called *Explorations*, which was produced and made by Zig Zag for Virgin. They wanted to get into content and make something they could distribute worldwide. The bumper said 'powered by Duracell.'

The arrival of digital has driven through ad-funded content. Much of brand-funded content on the web doesn't even play on TV, like Bebo-based drama *Kate Modern*, a groundbreaking online show which ran on the social network, stretching the boundaries of what online content providers and advertisers expect from the medium. A high degree of interactivity, serious traffic volumes and significant profits have made the whole industry sit up and take notice.

The concept of *Kate Modern* was created by Miles Beckett and Greg Goodfried to appeal to Bebo's core audience of 13 to 24 year olds. The story centred around the murder of Kate and her friends' attempts to find out who killed her. Four-minute episodes appeared six times a week in July 2007. On the final day of its first series a dozen episodes were shown in 12 hours with many viewers watching them as they were uploaded. The show averaged 1.5 million viewers a week.

Advertisers were offered a package including the number of video episodes their brand would feature in with more traditional banners around the screen. It attracted big sponsors like Proctor and Gamble and signed up Microsoft, Orange and Paramount amongst others. Bebo is now running four original series to capitalise on its success.

'I think there will be new methods of funding content, direct advertising, particularly online,' says Peter Bazalgette. 'I think the advertising sector has a very bright future online because it is now technically possible to target ads to individual tastes and pockets.'

The development of Internet Protocol TV to offer more targeted advertising is fascinating. 'If you live at number 47 and your best friend is at 51 on the same street and you are both watching *Coronation Street* but you have different habits and fashions then you could receive a different message,' says Nick Emery, chief strategy officer for WPP's pooled buying resource, Group M.

Paid for product placement regularly exists online and in US programmes from Manolo Blahniks in *Sex and the City* to Ford cars in *24* but is banned on British television though it is being considered as a viable source of funding. Ministers are consulting on whether to implement it as part of a European directive that would allow it in the

UK from 2010. Under proposals advertisers would pay for products featured in most TV programmes with the exception of news, current affairs and children's TV programmes. Media regulator Ofcom has moderately estimated that product placement would generate £25–£35 million after five years.

Brian Terkelson of Connectivetissue, the branded media content of US media agency MediaVest, argues that brands have made TV feel more real. As cultural artefacts they bring inherent story value. *Sex and the City* also featured Jimmy Choos, Apple and cosmetic brand Nars; in the US version of *The Office*, boss Michael Scott only uses Hewlett Packard. Mark Eaves, MD of Drum, the branded content arm of media agency PHD, claims that by legalising product placement in the UK and regulating it properly, 'you serve the viewer better.'

Transparency and viewer trust have become important issues in British TV following the spate of scandals that have rocked the TV industry. However, if an advertiser funds a programme they also have a role in shaping and determining the content. In *Sex and the City* a marketing campaign around Absolut Vodka was part of the story line, and in *Kate Modern* one of the characters worked for a PR agency, with Cadbury as a client with one of the characters dressed in a Creme Egg costume running around Leicester Square.

Another way of raising finance for the budget is to sell a show idea to different broadcasters in different countries. 'Co-productions help achieve a budget which would be unachievable with one broadcaster alone,' says Ben Barrett. 'They've been around for a long time but through necessity and financial reality in the UK, they are becoming increasingly used.'

Barrett's brief is to get different foreign broadcasters on board once a programme has been commissioned. 'Once we get money from a broadcaster, my job is to find the rest of the money to make the show happen. You bring in US and UK broadcasters who are interested in the same editorial. You often have to make two versions of the programme; we've done a number for five and National

Geographic. The show for Nat Geo is more simplistic, more scientific and filmed in high definition where as five would not be high definition and would have more human interest.

'We've also started doing a lot of co-productions with Canada as there are tax treaties between the UK and Canada, so the Canadian broadcasters can make a decent contribution to the budget if the budget is spent in that market. We did co-pro with five and Canada. The show was shot in UK but edited in Canada which makes a decent contribution to the Canadian economy.'

The art is in marrying up channels in different countries that have similar brands and audiences. In the UK five broadcasts many shows, such as *Ice Road Truckers* and *Deadliest Catch*, which also transmit in the US on male/science orientated channels like Discovery and National Geographic.

To survive, channels have to define their brand and have shows that are scheduled to attract and sustain viewers and therefore advertising revenue, and have a follow up audience who access the channels' spin off websites or discuss them in forums.

Channels spend huge amounts on research and planning programming strategies – it's all about identifying the audience demographic and directing the content and advertising accordingly. Each channel will have a target audience at its core – Bravo and five are unashamedly male, Living TV's is female, BBC 4 produces high-end documentaries in the arts, history and science, whereas BBC 3 is an intelligent youth channel aimed at 18–34 year olds with people in their mid 20s as it's 'sweet spot'.

And TV programmes themselves are increasingly becoming brands – a trend which is set to continue and expand. 'We don't come up with TV programmes anymore, we come up with brands,' says Kate Phillips. 'Every idea we come up with, we have to think, "How can we expand it? How can we spin it off? How can we make it bigger and better? What can we do with that brand?" '

For example, shows like *X Factor, The Apprentice, I'm A Celebrity* . . . and *Strictly Come Dancing* are not only on one night of the week but have several different spin-off shows on other sister channels and at different times. They also have websites, blogs and forums for discussions, products and in some cases web and board games as well as phone voting, all of which are huge additional sources of revenue.

'It's all about the programme as a brand and brand extension into other forms and other ways of delivering the enjoyable experience. The viewers don't just like the TV programme, they like the websites and the blogs and the games and the spin-off programmes,' says Fiona Chesterton who is a former commissioning editor and a former controller at the BBC and at Channel 4.

With content available on the net TV programmes are brands that are accessible at anytime a consumer demands – so they don't have to wait until they're scheduled on mainstream TV. 'You should be able to access a brand 24/7. At any point you should be able to go to a website or have a DVD or have a TV programme. You can access it in an instant, even if it not on the telly at that time. I think that's what it will really all be about in the future,' says Kate Phillips.

As a result schedules will be irrelevant because the viewer can choose what they watch, when and how. 'I think the future is non-scheduled,' says Kate. 'Scheduling to me is becoming less and less important.'

As are ratings – once the holy grail of broadcasters. Using the overnight figures collected from industry body, BARB, channels and programme makers would view these figures to see which programmes attracted the highest audiences, largest share of the available viewing audience and the number of people watching. These would determine which programmes were recommissioned as they attracted the most advertising revenue.

However, as the way people view programmes has changed, ratings are becoming less and less significant according to Kate. 'Ratings are really old fashioned now. In Children's BBC, we look more at what are they getting out of it, it doesn't have to be how many viewed the programme. It could be how many people went online afterwards or how many downloads to their phones they are getting and things like that. It's not just about ratings; it's about how best they are getting joined up about programmes by different means. Often, the ratings may not be hugely phenomenal but if you actually look at the people who love it and appreciate it, they will come back and the series will grow and grow into a brand.'

PROFILE

Ruth Wrigley, entertainment and format consultant, All3Media

BAFTA and EMMY award-winning Ruth Wrigley has more than 20 years' experience across a range of innovative genre-breaking programmes from serious documentary, comedy, and live studio entertainment shows to large scale multi-platform events. She is responsible for the first three BAFTA winning series of *Big Brother* in the UK and the multi-award winning *How Do You Solve A Problem Like Maria?* for BBC1.

Prior to *Big Brother*, Ruth played a pivotal role in creating and delivering the ground-breaking *Big Breakfast* for Channel 4. She has been head of entertainment and factual entertainment for Endemol, TalkbackThames, Leopard Films International, head of entertainment events at the BBC and was head of entertainment at Celador Productions.

April 2008 Entertainment and Format Consultant All3Media Group
Working on big entertainment projects across the whole group in Britain and internationally. The group currently consists of 15 companies, including Lion, Objective, North One and Maverick. I'm tasked with delivering the next big entertainment hit.

Nov 2007–April 2008 Head of Entertainment Celador Productions
Responsible for running a small team developing and devising
new factual entertainment and entertainment formats across all
channels, pitching and selling ideas to all key commissioners. Within
three months of arrival successfully secured a major new factual
entertainment series for BBC2, plus second series re-commissions for
series on BBC1 and Sky One, and *All Star Mr and Mrs* (ITV1)

2006/2007 Head of Entertainment Events BBC
Executive produced multi-award winning *How Do You Solve A Problem
Like Maria?* (BBC1) and consulted on US version (*You Are The One
That I Want*) and follow up series *Any Dream Will Do.* Oversaw and
revamped the *Eurovision Song Contest* and *Royal Variety Show.*
  Conceived and developed a number of new series including
When Will I Be Famous? with Graham Norton (BBC1).
  Ran Entertainment Events department to cover sick leave.

2004–2006 Head of Entertainment Leopard Films
Responsible for creating, developing, selling and delivering new
primetime entertainment and factual entertainment formats to all
broadcasters. Delivered primetime pilots to both BBC1 and ITV1
and won a commission for a primetime factual entertainment
series for Sky1.

2002–2004 Head of Factual Entertainment, TalkbackThames
Oversaw existing portfolio of shows including *This is Your Life*
(BBC1), *Wish You Were Here?* (ITV), *Farmer Wants a Wife* (ITV),
Gloria Hunniford Live (five). Ran development team, creating and
selling new formats to all broadcasters

Executive Producer
Distraction with Jimmy Carr (C4), *Boxing Academy* (five), *Zero
to Hero* (C4 education), *Design Wars* (ITV1), *Lifers Living with
Murder* (C4).

1998–2002 Head of Factual Entertainment Endemol UK
Started at Bazal as series producer on *We've Got Your Number*
(BBC 1). Executive producer for all Channel Four projects.
 Developed and ran the first three series of *Big Brother* in the UK.
 Devised and executive produced *Big Brother's Little Brother* (E4)
and streaming on E4 and the first series of *Celebrity Big Brother* for
Comic Relief (Channel 4). Developed and set up *The Salon* (C4).
Adapted and set up first series of *Fame Academy* (BBC 1) and *Spy
TV* (BBC1).

1996–1998 Executive Producer, Planet 24
Big Breakfast with Johnny Vaughan and Denise Van Outen
(Channel 4).

1995 Series producer, Hewland International
Wanted (Channel 4).

1994 Head of Outside Broadcasts Live TV
Responsible for overseeing large team producing 50 plus OBs
per week.

1992–1994 Series Editor, Planet 24
Big Breakfast with Chris Evans and Gaby Roslin.

**1998–1992 Series Producer/Producer Director/Studio and
multi-cam director, LWT Factual and Features**
*Weekend World, 6 O'Clock Show, London Programme, Telethon 1990,
South of Watford.*

1989–1992 Reporter and feature writer *Birmingham Post and Mail*

What does your job entail?
I am an executive producer or head of entertainment depending
on where I am. I work on big projects like something on the
scale of *Big Brother* or big Saturday night entertainment shows.

I specialise in innovative genre-breaking programmes from serious documentary, comedy, and live studio entertainment shows to large scale multi-platform events.

Did you always want to work in TV?
No. I wanted to be a journalist when I was at school, I was told I couldn't. I got a job on a local newspaper. I had no thoughts about going into TV. I went down to Fleet Street and hated it; it was full of chauvinistic men. So I got a job at Capital Radio as journalist/researcher, a Roger Cook-type who did investigative consumer stories. Then I applied for a job at London Weekend Television because they were looking for researchers who had a newspaper background and got it!

What was your first break?
It was on a local programme called the *6 O'Clock Show*. This was in the days when unions existed. They told me that actually there wasn't really a job. Another person had been issued a three-month contract which was illegal, so they had to give her the job but they liked me. They said that there might be another job soon and to apply again and I did and then I got it.

Did you have a game plan?
My fabulous career plan was to work somewhere that understood women with kids. My aim is to prove that you can work in telly if you have kids because I was told by my boss that I couldn't when my daughter was born. That was really unfair. This was going back over 20 years. I couldn't understand why, with all my experience, it was all or nothing. But my boss was probably 10 or 15 years older than me so was of a different generation. It's not the same anymore because my generation's at the top and they've come through a different route. I was with the last of the gin-and-tonic, wife-at-home brigade. It became my mission to succeed because I was told I couldn't.

How have you managed to combine motherhood and the long hours that TV demands?

I love my job but I never let it take over my life. It's different for each project, some projects demand very long hours, if they do fine, but I have a lot of time off afterwards. Others you can just do, they don't take over your life but they are quite intense when you're there. I've worked a four-day week since 1997. I refuse to do a five-day week. I've got kids so I do the job in four days. And I try not to work school holidays.

What's the most useful advice you've ever been given?

Peter Bazalgette (former chairman of Endemol) said always appoint people who you think are better than you because it only reflects well on you, and it does. Peter Bazalgette was my mentor. I have immense admiration for him because he's so clever but he's also a real person. He's hilarious, he likes eating and drinking, having a laugh and seeing his family and that's why I like him. He inspired me to realise you can make it in telly and be successful and have a life. He isn't threatened by anyone. He has complete self belief. I worked well with him because he just allowed me to do my thing but he was always there if there was a problem or if there was something I couldn't do; he dealt with it and that's what you want from a boss.

What skills do you have that contributed to your success?

I am really bossy and I have clarity of vision. I work well as a leader of a team because I'm clear about what I want, what needs to be done and then I enable people to do it because I don't do detail. I work really well with people who are very good at their specific thing because I let them get on with it. That's why I'm good on big projects because I take risks. You have to be brave, not just copying something that's already there. To do something new is such a struggle but I have some weird confidence where I think 'Look, I'm right, it will work' and most of the time it does. I suppose I'd be good

in a war! I've taken on problems on shows where there has been a
real crisis, where nobody else knows what to do. I do well in those
situations because no one argues with you! I'm a creative person
and when you believe in something and you've got a vision of what
you want to do, you'll get through because you never let it go.

What advice would you give someone starting out?
Be persistent when you're contacting someone for a job but don't
become a stalker. The hardest thing in telly is trying to get your
chance, to get your foot in the door. I would advise anyone to use
who they know rather than what they know to get in the door. But
once you're in, it's up to you as to what you make of it.

What's been your best moment so far?
The first ten minutes of the first *Celebrity Big Brother*. We had to
turn it around really quickly. One by one all the celebrities
dropped off and I was left with people I've never heard of and
someone from Boyzone! When we went on air it was traumatic, it
was minus fifteen degrees in the house and the toilet was broken.
It was just awful. I sat down with the series producer and said this is
either going to be the biggest load of shit ever and our careers are
over or its genius. Ten minutes later it's absolute genius. We did a
live nomination that night and that's when people realised that it
was really fascinating.

Any final words of wisdom?
It's only telly, no one dies. It's important and it's great when it
works but you're not operating, doing life-saving operations.
You're completely replaceable. It's recognising that fact. If you
drop dead tomorrow somebody else will just take your job.

What training issues does the industry face?
As well as being able to create and manage the creative process of
programme making in the UK, executive producers should be able

to advise producers in other countries on how to make them, when the programme is sold abroad.

Executive producers are expected to be able to create, produce, and sell formats that will work well overseas and be adaptable in different territories. However, after consultations the industry panel skills sector council, Skillset, found that not all EPs had the key skills required to sell their programmes abroad.

'A big area that came out as a real priority was helping creative thinkers with entrepreneurial skills,' says Alice Dudley of Skillset. 'There's a general recognition that people in executive positions needed help, particularly with the changing nature of the industry. Knowing how to exploit international sales and look at distribution. I think there is a big skills gap there.

'Research indicated that one of the biggest problems in TV was selling formats abroad and that senior execs weren't exploiting it quite enough and weren't looking at what they were producing and thinking of it as something that could be sold in a hundred different formats and distributed globally. They were just focusing on the smaller picture,' she adds.

The people who succeed in the industry and have long careers will be the ones to harness these entrepreneurial skills as selling formats globally will increase over the next few years. Peter Bazalgette predicts, 'The world wide format trend will continue to grow very rapidly because local content is the content that works. If you want a proven product you can't invent all your shows in one country.' So many countries buy formats from abroad but use 'local content', ie, their own celebrities, talent and contributors. For example *Dancing With The Stars*, *Secret Millionaire* and *Supernanny* have all been sold around the world and perform well in other territories.

What's next in training?
Editorial policy and guidelines have become tighter and in the light of huge fines and negative press attention producers have had to

improve their practices and be seen to be doing so and new training schemes will continue to address this.

'Editorial policy is hot at the moment,' says BBC talent executive Michelle Matherson. 'You would have to have some kind of briefing on this to work in TV.'

The BBC and Channel 4 have their own guidelines but Skillset is looking into formalising editorial guidelines by creating a 'passport' scheme of courses that will work across the TV industry.

'At the end of 2007 it became a huge priority,' says training manager Alice Dudley. 'We were asked by Ofcom to look at the skills issues there and try to come up with something which addressed this.'

As budgets get tighter and programme genres fuse, there is also a trend towards multi-skilling which requires training. It is no longer enough to be able to perform just one job or one role. As programmes diversify the skills they require become more diverse and as budgets get smaller one person has to do more than one job.

'Anybody aged 22 can shoot and edit their own stuff,' says Peter Bazalgette, who in his role as deputy chairman of the National Film and Television School has doubled the size of the student intake, built a new building, made the funding basis more diverse and put in new courses. 'I'm not very good at that, the next generation is. It's definitely a sign, neither good nor bad, absolutely inescapable and inevitable with the onward march of technology,' he says Peter Bazalgette.

It is not enough to be just a researcher or AP. Researchers, APs and producers who can also shoot a camera themselves are already in demand and the trend will continue. 'They have to learn those skills; if they do they'll be at a huge advantage going in for an interview,' says Richard Hopkins.

And as well as self shooting, being able to understand and record sound, and in some cases edit your own material are trends which continue. In the past, these jobs were all done by different people – and still are – but the demand and availability of multi-skilled TV producers will continue to grow. In the United States these multi-talented creatives are called 'preditors' because they can produce and edit.

'Desktop editing has become a big thing,' says Richard Hopkins. 'It saves us a lot of money, and money is becoming tighter and tighter so it will become more of a necessity. When we're cutting an early important tape on a desktop, we don't call in the actual editor until the last minute because we want to save money. So any skills in that area are very useful.'

However, although the TV industry is driving these changes, it's down to the freelancer to find and fund the cost of formal training themselves. 'It's not really our problem,' says Richard Hopkins of independent production company Fever Media. 'If they have those skills when they apply then it's an advantage. If we were a bigger company, we would start to structure training for runners and the like but because we're a small company, we just pluck the best people out of the freelance community. Desktop editing is not complicated. I think most people nowadays would have a good sense of it, you can get it on Mac now and it's not that difficult a skill to acquire.'

Skillset's TV Skills Council meets annually to discuss industry concerns and has identified multi-skilling as an issue that needs to be addressed by increased, affordable training. It allocates funds for schemes and accredits training schemes from outside companies to ensure quality control.

'There's a cross over between documentaries, between factual and drama,' says training manager Alice Dudley. 'In factual programming drama is used to illustrate points in documentaries. Often the drama bits in that look quite weak and so you needed a director who can do both. That kind of multi-skilling thing is a huge priority. Obviously it's better to have somebody who can do everything or to have just one person rather than two.'

What are the next big ideas?

'Television goes in fits and starts,' says Peter Bazalgette, who created pioneering new genres with *Changing Rooms, Ground Force* and gave birth to the celebrity chef in *Food and Drink*. 'It hits the public and business very unpredictably. I remember people saying to me in the early '90s, "It's all been done, there's nothing new in television". And indeed between 1945 and 1995, TV was dominated by about five genres. They were all invented by the Americans after the war: game shows, sit-com, sport, news, animation.

'But actually between '95 and 2005 a new generation emerged. *Who Wants To Be A Millionaire?* reinvented television, a game show during prime time. *Survivor* and *Big Brother* created a whole new genre of reality shows. Were they games shows? Were they human-interest shows? The second half of the '90s were incredibly revolutionary and productive for television genres.'

Bazalgette believes instead of one big thing, the trend will be for ideas to evolve and emerge from existing ideas. 'The "noughties" by comparison, have been quite tame, we've been perfecting and coming up with variations on these themes. There will be more break-throughs at some point but you can't predict when.' TV will continue to adapt and rip off successful shows, adding a twist and something new to keep the ideas fresh and pushing things forward.

'How do you stay one step ahead and predict or create future trends?' asks Conrad Green. 'It sounds obvious, but watch a lot of TV.'

Richard Hopkins agrees, 'TV is very cyclical. Commissioners tend to swim around, like shoaling fish, into the same areas at the same time and feed those areas with lots of ideas and then all swim off in another direction, desperate for the next hit show. A show works in one genre and then everyone tries to think of how they can do that show in a different way. So it's very difficult to say what's going to dominate the next few years. But I think humiliation reality will become more niche. Sooner or later there will be a big new quiz show that will inspire lots of imitators, like *Deal Or No Deal.*'

'Nobody can really predict future trends,' says Fiona Chesterton. 'Nobody predicted reality TV before it happened. It may have evolved from the soaps, in some ways you can say they had elements of reality shows.'

'A lot of people said that live entertainment shows were dead but five years later we get *Strictly Come Dancing.* One thing you can say is you just can't predict where the next big thing is coming from!'

TV will continue to evolve with existing formats leading the way. They'll be re-packaged and adapted to drive popular genres forward. For example, just as *Strictly Come Dancing's* format was the inspiration for other BBC shows, *How Do You Solve A Problem Like Maria?*, *Maestro*, and *The Choir*, the trend is by no means exhausted. 'We do a lot of talent shows but I think talent isn't going to go away now,' says the BBC's Kate Phillips. 'That sort of feel-good reality has not seen its last yet; the sort of *Strictly, Maria, Joseph* reality that is tough and hard work but makes people look and feel good at the end of it.'

'Look at the figures for *Britain's Got Talent,*' says ITV Entertainment's Katie Rawcliffe. 'I think in some ways we're going back to old variety. ITV put *Dancing on Ice* on a Sunday. Variety entertainment is really big.'

Big Brother was launched in the UK in 2000 but is still on air and a successful brand and its tentacles are still far reaching. The demand for reality shows looks set to continue. 'Reality has been around for years and still going and I don't think it's going anywhere,' says Rawcliffe. 'People always say reality programming is terrible but they still really buy into it.'

'Reality will never be dead. There will always be a post reality, there will always be post talent shows. It's just how you do it differently,' says Kate Phillips. 'How many shows are there with a panel of judges and voting off, elimination? It's not the subject area it's more about the mechanic of how you do the programme so it doesn't seem derivative.'

TV is a medium which eats itself and reproduces the content in different ways, borrowing and stealing elements of hit shows and adding new ones. 'Reality brands will keep developing. People will keep working on those, sort of expanding and enhancing those formats,' says Katie Rawcliffe.

'They'll be a big new reality show that harnesses 360-degree content, that will make *Big Brother* suddenly seem quite old fashioned,' says Richard Hopkins.

TV will change in how it produces and is seen to make and manipulate programmes. In the light of several TV scandals, including dubious practice with phone voting, the Queengate scandal where footage of the Queen was shown to have been edited and taken out of context, TV has to clean up its act to win back viewer trust.

'Factual entertainment is going to go through a difficult time because a lot of the factual entertainment formats that we know and love now, if they are scrutinised, do tend to exaggerate their truths,' says Richard Hopkins. 'I don't know if the heat will suddenly go off that. Anything from *Ramsay's Kitchen Nightmares* to *Wife Swap* to *Supernanny* will become difficult to deliver. Formats like that massage reality, they are formatted in a way that's a condensed version of real life, that allows producers to exaggerate real life and that's what makes them entertaining.'

With a TV producers now have direct access to the real lives and the talents of undiscovered characters – from musicians, comics, presenters and experts who now have the benefit of using the internet as a platform to showcase their abilities. Producers will profit from this free access and be able to exploit this wealth of undiscovered talent.

'I think that far from competing with television, Youtube and the like are fantastic for developing the next generation of talent, and allowing them to advertise their wares and allowing the more contextual media to find that and invest in that talent,' says Peter Bazalgette. 'The next generation of writers, directors and comedians are all there, in their bedrooms, messing about with video and they will be spotted in the future.' This wealth of talent will become the new TV stars. It has never been easier to get on TV.

PROFILE

**Anna Richardson, TV presenter, executive
producer and format creator**

Anna Richardson is a TV presenter, series
producer and executive producer in
development. Anna started her TV life in the
heyday of the *Big Breakfast* – Channel 4's most
successful breakfast show. She was soon spotted
for her wit and frank down-to-earth style, and
beat some of the best-known names in British
TV to present the cult teen series *Love Bites* for
ITV. The series was so successful it won awards

in the UK and America, and Anna was nominated for an RTS award.

For the next six years she hosted a number of series for the BBC
and ITV, including *Love Bites Back, Dream Ticket, Des Res, Maternity
Hospital* and ITV's flagship film series *Big Screen,* where she met
and interviewed some of the biggest names in Hollywood.

Anna turned her attention to writing and producing successful
television formats for the BBC, ITV and Channel 4, including
No Waste Like Home, Turn Back Your Body Clock, and Channel 4's hit
series *You Are What You Eat* with Dr Gillian McKeith, which has
subsequently been sold to over 40 countries.

She is a Channel 4 presenter hosting prime time series *The Sex
Education Show* and *Supersize v Superskinny.*

How did you get your first break?
I did lots of work experience at BBC North in Manchester, for
Reportage and Network 7 but I just couldn't get a break as a
researcher. I was getting so fed up with it that I decided to do a
post grad course in magazine journalism. I moved to London and
then I saw an ad in *The Guardian* for *The Big Breakfast* for a day
researcher. There were 200 people who went for the job and
I got it. The job was literally a baptism of fire. It was a 24/7, all
consuming job but it was the best training for me. I did six months

there but it was exhausting both emotionally and physically. After I finished there I had 12 weeks off to recover!

Did you have a game plan?
I have been determined to work in TV ever since I was seven years old. I used to drive my family mad with it! I knew I would present and I knew I would produce. My game plan was that I refused to start as a runner. I wasn't going to be making tea for anyone, which is why I applied for researcher jobs and did my post grad.

How would you describe yourself?
Honest, organised. I hope I'm inspirational, fun. I acknowledge that I am confrontational, opinionated and controlling!

What does your job involve on a day to day basis?
At the moment I am working as the presenter of *The Sex Education Show* on C4 so I work at least four days a week for 12 hours a day. It is a magazine format so one minute I may be subjected to a fertility test, the next minute watching a live birth, and the next interviewing teens on their opinions on pornography.

It's fascinating. I go on journeys myself too, therefore it's very personal and challenging. It's not just straightforward presenting as the programme covers quite private stuff. One my strengths, but also a huge failing, is that I am too honest and open. It is entertaining for the viewer but can cause upset at home. However, I always wanted to be transparent.

How did you get your first break as a presenter?
Jane Root at Wall to Wall spotted me as potential for on-screen talent. I was working as an AP there and they were looking for a reporter to work with Lorraine Kelly on a programme called *Ultimate Shopping Guide* so I decided to give it a go. Jane said I had a real talent for it and I absolutely loved it.

My real lucky break with TV presenting came from a man called Conrad Green, he is now a major EP in LA on programmes like

American Idol and *Dancing with the Stars*. He was working at ITV and they were looking for a presenter for a Saturday morning show called *Love Bites*. It launched my presenting career.

How long have you been a presenter for?
I was a full-time presenter continuously for six years, up until 2000. I did everything from live, studio, location and audience, and I was doing a programme called *Maternity Hospital* for the BBC in Manchester. I then went on to present a programme called *Big Screen*, which was an ITV weekly film show for about four or five years. It was amazing, I got to interview celebs, film stars and watch private viewings. They were the best times of my life. I got to go to Cannes every year, too. When this ended I just couldn't get a job on screen for about 18 months. It was so frustrating because I had an agent and it is what I wanted to do! There was a shift in TV; everyone wanted experts, on history, nutrition, beauty, property.

What did you decide to do?
I decided to snap myself out of it and get back into production. No one would touch me for a while because they found it strange that I was going from being a presenter. It was really hard to get back in. Then Damon Pattison, who was head of development at Celador, took me on as a development producer. I had the most fantastic three years with him. We got our big break together as we co-created *You Are What You Eat* and found Gillian McKeith. We then sold this to Channel 4. I was extremely happy and I was promoted to development executive. After this I left Celador and ran development departments for other companies. I secured a commission with ITV with talent and the format of a programme called *Sally Morgan, Star Psychic*. It was a 10 × 60' series and was my first job as a series producer and co-executive producer. I then series produced a 6 × 60' series for five called *My Body Hell*.

I moved on to become an executive producer at Shine running the features development team. Then out of the blue

Channel 4 approached me and asked me to present on *Supersize vs Superskinny*. Then off the back of that I was given my own show *The Sex Education Show*. Channel 4 has been very good to me.

How have you sustained a career as a presenter?
I'm not sure I have! People can think you're good and hot one minute but the moment of glory will always pass. Then you have to decide what to do next.

What advice would you give someone wanting to be a presenter?
Have a back up plan. Don't rely on it as a full time job as it is so hard to get into. You need to learn to rely on your other skills too. Having production as a background helps so much.

What's more enjoyable – presenting or producing?
They are equally enjoyable and difficult in different ways. Producing is an excruciatingly difficult job. There's never enough money or time, and it's very stressful because you need to deliver fantastically. Presenting is exhausting in a different way. You always have to be very nice and the grin definitely wears thin. You always need to listen carefully and be prepared with an answer. Presenting is a lot about being yourself and that really isn't always easy! It can also be very personal. People always judge you on whether you're fat or thin, good looking or not and your mannerisms which irritate them. Fifty per cent say you look great but there's always the fifty per cent who say you don't.

I went on forums on the internet but stopped when someone said I was a fat slag. It was when I was on *Supersize vs Superskinny* and someone had said that with all the diets I was on I hadn't lost any weight. You're so exposed to people's madness.

Why did you choose to focus on working in development?
I trained as a journalist so ideas and writing were in the blood. Damon gave me a break and I happened to be quite good at it. I'm an ideas and writing person.

How do you come up with ideas?

Expose yourself 24/7 to radio, magazines, newspapers, and books, and especially BBC Radio 4 in my case. As long as you meet interesting people you come up with ideas. Sometimes you can be dreaming and it will spark an idea. Sometimes you could be buying ice-cream and meet someone with the most fascinating life story. Talent spotting is good for ideas. I found Gillian McKeith and came up with *You Are What You Eat*.

How did you come up with *You Are What You Eat?*

I called agent Nicole Ibison from First Artists Management to see if they had any nutritionists. I met Gillian and she was so energetic I knew she would be great on TV. The first thing she said was, 'What is your poo like?!' With TV talent you're always looking for the next big thing in personality. It's what producers call a 'water cooler' moment. Something that is a big and shocking expose sells. Like Gordon Ramsay with his swearing and Gillian talking about poo! Often when you find characters that are larger than life you know that even if they are loved or hated they will always bring in ratings.

How do you sell an idea?

I'm passionate about development. People disregard those who work in development and think they're failed production people, but without development there are no programmes! Production teams should kiss their feet and cover them in rose petals! It's your job to have a good relationship with commissioning editors. It's all about relationships. TV is like one big marriage!

What achievement are you most proud of?

My breakthrough point was creating *You Are What You Eat*. It was a massive success. Once I was out to dinner and there was a whole group of women discussing it, saying did you see that new programme last night, wasn't it great? I had to go over and say, I came up with that!

What's the best thing about working in television?
What other job would allow you to meet film stars, travel the world, cry with ordinary people and work with some of the most creative minds? It is outstandingly fulfilling.

What's the worst thing you ever had to do working in TV?
During the filming of *Sex Education* I had to do a live fertility test on camera and found out I had low fertility rates which was hard. I then had to go straight to do a shoot to learn how to be a burlesque dancer. I had also eaten something funny for lunch and I couldn't stop being sick.

What's the best piece of advice you were given?
Danielle Lux, who is Managing Director of Celador, said 'Anna, there are very few nasty people in TV. Just people in the wrong jobs.' It has stayed with me as it's so right. People are good people but in the wrong job you can be an absolute maniac.

What advice would you give someone wanting to break into TV?
Put a smile on your face and just say yes!!

What personal qualities have helped you get on?
A lot of my personality is a double-edged sword but honesty, openness, humour and optimism.

What will you be doing in 10 years' time?
I'd like to think I had come up with the next *Who Wants to be a Millionaire?* and relaxing on a yacht somewhere!

What's the best way to get ahead in television?
Dedication and a lot of alcohol!

An A to Z of contributors and their credits

Alice Dudley, training manager at Skillset

Andrew O'Connor, chief executive officer, Objective Productions

Andrew Chaplin, producer/director and founder of Headhunt TV

Alex Marengo, specialist factual series producer. Credits include *I Shouldn't Be Alive* (Discovery US/Channel 4), *We Built This City – New York* (Discovery Channel/five), *Ancient Superweapons – The Ram, The Claw and The City Destroyer* (Discovery)

Anna Blue, entertainment series producer. Credits include *Ant & Dec's Saturday Night Takeaway* (ITV1), *The Wish List* (ITV1), *The All Star Cup* (ITV1), *Tim Lovejoy and The AllStars* (Sky1), *The Games* (E4), *Big Brother* and *Big Brother's Little Brother* (C4)

Anna Keel, self shooting producer/director. Credits include *The Apprentice* (BBC 1), *Supernanny* (Channel 4), *Air Medics* (BBC1)

Anna Richardson, Channel 4 TV presenter and executive producer in development, format creator. She devised *You Are What You Eat* and *Turn Back Your Body Clock*; presents *The Sex Education Show* and *Supersize vs Superskinny*

Ben Barrett, commercial director, Zig Zag Productions

Charles Martin, drama director. Credits include *The Giblet Boys* (ITV1), *My Life As A Popat*, *The Sara Jane Adventures*, *Skins* (C4)

Conrad Green, British executive producer based in Los Angeles. Credits include *Dancing With The Stars* (ABC Network USA), *American Idol*, *Diners* (BBC 3), *The Murder Game* (BBC 1), *Celebdaq* (BBC2/3), *The Bunker: Crisis Command* (BBC2), *Big Brother* (Channel 4). Series producer *Popstars*

Claire Richards, self shooting assistant producer. Credits include *Would Like To Meet–Again* (BBC 2), *Manhunt 1* and *2* (ITV1), *Great Food Live* (UKTV), *Jade's PA* (Living)

Daisy Goodwin, managing director of Silver River TV. Credits include *Natural Born Sellers, House Doctor, Jamie's Kitchen, Grand Designs, How Clean Is Your House?, Would Like to Meet, Your Money Or Your Life, Property Ladder, The Sex Inspectors*

David Minchin, self shooting researcher/development researcher. Credits include *Accidental Heroes* (BBC1), *The Boss Project* (five Pilot), *Wish You Were Here Now* & *Then* (ITV1), *The British Soap Awards* (ITV2), *WAGs Boutique* (ITV2), *The X Factor* (ITV1)

Deborah Kidd, producer/development producer at CTVC, IWC and Darlow Smithson Productions. Credits include *The Insider* (Channel 4), *The Root of All Evil* (Channel 4), *Ancient Superweapons* (Discovery/five), *Crashes That Changed Flying* (WNET/five)

Dominic Crofts, assistant producer. Credits include *The Alan Titchmarsh Show* (ITV1), *The Match* (Sky One), *Rory & Paddy vs the UK* (five), *David Beckham Documentary* (19 Management), *Comic Relief Does Fame Academy On BBC3* (BBC3), *Diary Room Uncut* (E4)

Eiran Jones, careers information co-ordinator at Skillset

Emily Gale, head of talent at TalkbackThames

Emily Shanklin, talent manager at Betty

Fiona Chesterton, former director of television, Skillset, now a consultant to Skillset

Glen Barnard, runner. Credits include *The Sex Education Show* (Channel 4), *Watchdog* (BBC1), *Rogue Traders* (BBC), *I'm A Celebrity . . . Get Me Out of Here!* (ITV2)

Grant Mansfield, chairman, RDF Media Group Content, Managing Director, RDF Television. Oversees *The Secret Millionaire* (Channel 4), *Wife Swap* (Channel 4), *Shipwrecked* (Channel 4), *Dickinson's Real Deal* (ITV1), *Oz and James's Big Wine Adventure* (BBC), *Scrapheap Challenge* (Channel 4), *Ladette to Lady* (ITV1), *Personal Services Required* (Channel 4), *Don't Forget The Lyrics* (Sky One)

Harjeet Chhokar, assistant producer/development AP. Credits include *Big Brother 8, Wife Swap 9, Big Brother Celebrity Hijack, The Secret Millionaire* 3 (all Channel 4)

Helen Beaumont, runner. Credits include *Gladiators* (Sky One), *The One and Only* (BBC), *The Sex Education Show* (Channel 4)

Jenny Popplewell, shooting AP/development AP. Credits include *No Place Like Home* (ITV1), *Bad Boy Racers* (five), *Change The Day You Die* (Sky One)

Jo Taylor, talent manager, 4 Talent, Channel 4. Credits include head of production talent, Optomen Television, BBC TV managing editor documentaries and contemporary factual, BBC TV production executive general factual

John Adams, researcher. Credits include *Katie and Peter Unleashed* (ITV2), *That Antony Cotton Show* (ITV1), *Let Me Entertain You* (BBC 2), *Alan Carr's Celebrity Ding Dong* (Channel 4), *The New Paul O'Grady Show* (Channel 4)

Jon Crisp, self shooting AP/PD. Credits include *Bad Boy Racers* (five), *Desperate Virgins* (Channel 4), *Freaky Eaters* (BBC 3)

Julia Waring, head of creative resources RDF Media

Kate Phillips, head of BBC Entertainment Development

Katie Rawcliffe, head of entertainment series, ITV1. Credits include *Dancing on Ice* (ITV1), *Hell's Kitchen* (ITV1), *Holly & Fearne Go Dating* (ITV1), *Comic Relief Does Fame Academy* (BBC1), *I'm A Celebrity . . . Get Me Out Of Here!* (ITV2), *The Match* (Sky One)

Liz Mills, founder and managing director of Top TV Academy. Former head of reality at Endemol. Credits include *Pet Rescue* (Channel 4) and *Crimewatch* (BBC),

Louise Mason, edit producer/producer director. Credits include *Kill It, Cook It, Eat It* series 3 (BBC3), *Dog Borstal* (BBC3), *Shipwrecked* (Channel 4), *Would You Buy A House With A Stranger?* (BBC3)

Lucy Reese, media lecturer Westminster Kingsway and City of Westminster Colleges and series producer. Credits include *Killer Facts* (Discovery), *Greatest Ever Movies* (five), *Greatest TV Moments* Series 1 (five), *Greatest TV Moments* Series 2 (five), *50 Worst Decisions of All Time* (Sky One)

Matt Born, managing director, DV Talent

Michelle Matherson, talent executive, BBC Factual

Moray Coulter, production and talent executive for factual entertainment ITV

Nick Holt, self shooting observational documentary producer/director. Credits include *Guys and Dolls* (five), *The Mentalists* (five), *90 Naps a Day* (Channel 4), *Storm Junkies* (Channel 4)

Peter Bazalgette, former chief executive Endemol, deputy chairman of the National Film and Television School, former independent producer and format creator. Regarded as the founding father of reality television

Puja Verma, runner. Credits include *The Girl with 8 Limbs* (Channel 4), *Relentless* (Channel 4)

Rachel Roberts, broadcaster and freelance journalist. Writes for women's magazines and national newspapers including *The Observer* and *The Daily Mail*

Ruth Wrigley, entertainment and format consultant at the All3Media Group. Credits include *All Star Mr and Mrs* (ITV1), *Big Brother* (Channel 4), *How Do You Solve A Problem Like Maria?* (BBC1), *The Big Breakfast* (Channel 4)

Richard Drew, US based British executive producer. Founded his own New York-based production company, Savannah Media. British programme credits include *Big Brother* (Channel 4), *Survivor* (ITV1), *Fashanu's Football Challenge* (ITV1), *The Club* (ITV1), *The Big Breakfast* (Channel 4)

Richard Hopkins, managing director of Fever Media. Credits include *Strictly Come Dancing* (BBC1), *Big Brother* (Channel 4), *The 11 O'Clock Show* (Channel 4), *The Big Breakfast* (Channel 4)

Simon Warrington, assistant producer. Credits include *Big Brother 8* (Channel 4), *Britain's Greatest Managers & Football – The Glory Years* (Bravo), *Strictly Come Dancing – It Takes Two* (BBC2), *The Match* (Sky One)

Sue Dulay, series editor at Ricochet. Credits include *Selling Houses* (Channel 4), *Breaking Into Tesco* (five), *Open House* (BBC 1)

Susannah Haley, researcher/AP and script reader. Credits include *The Charlotte Church Show* (Channel 4), *Russell Brand's Big Brother's Big Mouth* (Channel 4) and *Big Brother's Little Brother* (E4), *I'm A Celebrity . . . Get Me Out Of Here!* (ITV2)

Tim Hincks, chief executive officer Endemol UK. Format creator as creative director of Endemol working across the company's output

Victoria Ashbourne, executive producer. Credits include *My Little Soldier* (ITV1), *All Star Mr and Mrs* (ITV1), *The Dame Edna Treatment* (ITV1), *The New Paul O'Grady Show* (ITV1)

Useful TV terms

Below is a short, selective list of terms and phrases which will help you navigate your way through this book and your first forays into TV.

Assistant producer (AP). The next rung up the ladder above a researcher and before a producer. Probably one of the most diverse job roles in the industry which varies according to the genre of programme.

Audio. Sound.

Beta SP. Digital tape format of broadcast quality.

Bumper. The opening and closing graphics before and after ad breaks, known as 'break bumpers'.

Close-up. Shot showing a person's whole head from the shoulders upwards. Close shot of an object

Cue. A signal for action.

Cut. To terminate a programme, image or scene abruptly; to make a sudden and complete change from one to another.

Cutaway (C/A). A short shot used to illustrate someone's character and cover a cut in an interview.

Commissioning editor. Responsible for commissioning programmes from independent production companies and overseeing the programme while it's being made. Often specialises in a genre, for example comedy, entertainment or factual.

Docu-drama/Drama-doc. A form of documentary based on historical events but includes actors reconstructing real events dramatically.

Documentary. Filmed or videotaped stories that are based on facts, people or events.

Dub, Dubbing. To copy by playing back on one machine and recording on another. Sound dub mixing fuses the commentary with the sound effects and music.

DV digital video/Domestic video. Used professionally as in Panasonic's DVC-Pro, and Sony's DV, and DV Cam.

DVC. Light-weight digital video camera often used to shoot programmes and used by shooting researchers, assistant producers and producer/directors. Usually hand held such as PD150, Sony 27.

Edit. A generic term where the raw material is edited to create a master programme. To link one piece of audio or videotape to another, or to create a master tape of an audio or video programme, usually from a variety of source media.

EDL (edit decision list). A computer program that allows the user to re-create or modify an audio or video programme.

Effects (Sound). Recorded noise other than music or speech.

Establisher. Usually a wide angle shot that establishes a location, its contents and characters.

Exterior. Any out-of-doors shooting.

Eye-line. The direction a person on camera is looking.

Executive producer. Top of the creative ladder. Responsible for quality control in terms of content, contributors, vision, look and style. Often oversees more than one project at a time, they manage the relationship with the commissioning editor at the broadcaster and also sell programme ideas to UK and foreign broadcasters.

GVs. General views. This is a geography shot that shows the whole location (as much as is practial) or the location in its setting. It's the GV that establishes where you are. Also known as an establisher.

HD. High definition.

HDTV. High definition television. One of the coming standards in the future of television technology. A TV format capable of displaying on a larger screen and at higher resolution.

Jump cut. A jarring edit that can result from shifts in camera angle, frame size, editing commentary. An edit on the same subject that doesn't work.

Multimedia. The delivery of information, via personal computer or interactive player, that combines text, graphics, audio, still images, animation, motion video from a CD or DVD, magnetic disk, optical disc, video or audio tape.

NTSC. Video format used in America

Offline. An editing suite with a low resolution computer and disk-based edit system in which the creative editing decisions can be made at lower cost and often with greater flexibility than in an expensive fully equipped online suite.

Online. An editing system where the programme master is created. An on-line bay usually consists of an editing computer, vision mixer, audio mixer, one or more channels of DVE, character generator for captions, and digital video tape machines.

Producer. A producer is normally in charge of a big part show, either an episode of a long-running series or a specific part of a larger show. They are responsible for content and managing a team to deliver it. On a studio show they would also write the script and oversee the casting of contributors.

Producer director (PD). In charge of one episode of a show within a series, for example a one hour stand-alone documentary or inserts for a news or magazine show. They plan it, write the script, find and select contributors, film, edit and deliver the show.

Radio mic. Transmitter or wireless microphone connected to a small radio transmitter, used in situations where cables would be difficult or impossible to use.

Raw tape. A term sometimes used to describe tape that has not been recorded. Also called virgin or blank stock.

Researcher. Responsible for finding stories, contributors, locations, clips and archive and storylines. Finds ideas, writes briefs, interviews, questions and scripts. Usually works with an AP or producer.

Runner. The first paid entry level job in TV. The role varies from company to company but the job usually involves running errands.

Rushes. The raw unedited material shot for the programme.

Series producer. Responsible for managing the team, the budget, the schedule, the content and the edit for a series. Taking the series and the production from the intial idea and format, overseeing it in the edit and delivering it as a finished show. Reports to the executive producer.

Stock. The tapes for the camera that will be used to record the programme.

Sync. Speech/interview comments used in a programme.

Time code. A digitally encoded signal that is recorded on videotape to identify each frame of video by hour, minute, second and frame number.

Titles. Also known as opening titles that appear at the beginning of a programme.

Tripod. A three-legged stand on top of which a camera is mounted.

VT inserts. Edited, packaged features that normally run for up to five minutes within a larger programme or alongside other items.

Zoom. To gradually change the field of view of a camera lens from wide to narrow angle (zoom in) or narrow to wide angle (zoom out).

Index